Integrated Water Resource Planning

Integrated Water Resource Planning provides practical, evidence-based guidance on water resource planning. In a time of heightened awareness of ecosystem needs, climate change, and increasing and conflicting demands on resources, water professionals and decision-makers around the world are on a steep learning curve. This book presents an international examination of water reform experiences, and provides lessons in how to manage environmental uncertainties, long term management, and increase in demand. It breaks the process down into a series of common steps, applies programme logic and evaluation theory, and discusses best practices in assessment, decision making and community engagement. Importantly, it recognizes the large variation in available knowledge and capacity, risk and scale, and discusses a range of approaches that can be used for different circumstances.

The book will fill in the gaps for professionals in interdisciplinary teams including sociologists, hydrologists, engineers, ecologists, and community consultation specialists, by providing a basic grounding in areas outside their usual expertise, and will provide ammunition to community stakeholders in their quest to ensure that water planning outcomes are justified and justifiable.

Case studies provide an understanding of the context, practical tools and implementation techniques for achieving sustainable outcomes, and the multi-disciplinary approach and insights offered in this book will be transposable and instructive for water professionals worldwide.

Claudia Baldwin lectures in regional and urban planning at the University of Sunshine Coast, researching in social, institutional issues, governance and collaborative processes in water, coastal and land-use planning and management. Where possible, she uses interactive and visual research techniques. She has more than 30 years' experience in policy, planning and management positions in the Queensland government, Great Barrier Reef Marine Park Authority, and as a consultant overseas and in Australia.

Mark Hamstead is a respected water resource policy and management specialist with 29 years' experience. He provides advice and analysis to government agencies and the private sector. His areas of expertise include water policy and legislation, water planning and management, water entitlements, water trading, water accounting, water resource assessment and water usage metering and monitoring. In recent years Mark has prepared, or been a major contributor to, a range of reports for Australian water authorities including such things as water planning, water trading and managing water for environmental outcomes. Prior to becoming a consultant in 2005 Mark worked for many years in government water agencies in Australia.

Integrated Water Resource Planning

Achieving sustainable outcomes

Claudia Baldwin and Mark Hamstead

Routledge
Taylor & Francis Group

LONDON AND NEW YORK

earthscan
from Routledge

First published 2015 by Routledge

2 Park Square, Milton Park, Abingdon, Oxfordshire OX14 4RN
711 Third Avenue, New York, NY 10017

Routledge is an imprint of the Taylor & Francis Group, an informa business

First issued in paperback 2018

British Library Cataloguing in Publication Data
A catalogue record for this book is available from the British Library

Library of Congress Cataloging-in-Publication Data
Baldwin, Claudia (Claudia Lilian)
Integrated water resource planning : achieving sustainable outcomes /
Claudia Baldwin and Mark Hamstead.
pages cm
Includes bibliographical references and index.
1. Water resources development--Planning. 2. Water-supply--Planning.
3. Water quality management. I. Hamstead, Mark. II. Title.
HD1691.B355 2014
333.91--dc23
2014004585

ISBN13: 978-0-415-83548-0 (hbk)
ISBN13: 978-1-138-37259-7 (pbk)

Typeset in Garamond by Fakenham Prepress Solutions, Fakenham,
Norfolk NR21 8NN

Contents

List of figures

List of plate figures

List of tables

List of boxes

List of acronyms

ABS	Australian Bureau of Statistics
CMA	Catchment Management Agencies, South Africa
CTP	cease to pump
DWA	Department of Water Affairs, Republic of South Africa
EU WFD	European Union Water Framework Directive
GL	gigalitre
GWP	Global Water Partnership
IWRM	Integrated water resource management
MCA	multi-criteria analysis
MDBA	Murray Darling Basin Authority, Australia
ML/d	megalitres per day
NGO	non governmental organisation
NRM	Natural Resource Management
NSW	New South Wales, Australia
NT	Northern Territory, Australia
NWA	*National Water Act 1998*, South Africa
NWC	National Water Commission, Australia
NWRS	National Water Resources Strategy, South Africa
NWI	National Water Initiative, Australia
RBMP	River Basin Management Plan
SEQ	South East Queensland
UN	United Nations
UNDP	United Nations Development Program
UNEP	United Nations Environment Program
USA	United States of America
WAP	water allocation plan
WFD	Water Framework Directive
WSP	water sharing plan
WSSD	World Summit on Sustainable Development
WUA	water user association

Foreword

With ever growing demands for water due to expanding cities, agriculture, energy, industry and mining, as well as more recognition of the benefits that arise from maintaining water-dependent ecosystems, integrated and sustainable water management is an increasingly challenging issue. Nations across the world are grappling with difficult decisions about how much water to allocate and in some cases how to scale it back. While integrated water resource planning and management is an internationally accepted paradigm, the practice of implementing it is in its infancy.

This book is a timely response. It fills a gap by providing guidance in policy and practice of water resource planning. It builds on the authors' own experience by drawing from examples around the world. While the authors do not claim to cover all issues in all regions, many of the cases resonate with my own experience in the STRIVER project[1] and highlight the importance of the science-policy-stakeholder interface in water management (Gooch and Stalnacke 2010). Most importantly, the book includes a range of strategies applicable to developed and developing countries, for addressing the many challenges we face.

The book derives from both authors' understanding of government policy development and the intricacies of its on-ground application. However, they also have skills and experience that complement each other. Claudia Baldwin, an environmental planner with a social science background, contributes institutional insight, consultation practice, cross-disciplinary interaction, and academic oversight. Mark Hamstead, a consultant and former water agency staffer with a hydrology/engineering background, brings technical expertise about hydrological systems, program logic, evaluation, modelling and scientific assessments as well as solid experience in the many practical challenges of water planning and management. Their shared values for a reasoned and sound process as a basis for achieving fair outcomes permeate the book.

---

1 STRIVER: Strategy and methodology for improved IWRM – An integrated interdisciplinary assessment in four twinning river basins (2006–2009), is a project supported by the European Commission Sixth Framework Programme. See http://kvina.niva.no/striver/

The past ten or so years have seen nations and their water planners tackling water reform, grappling with institutional reorganisation, limitations in capacity, more informed stakeholders, with never enough science or resources to do justice to a critical issue. Food security, health, and economic development all rely on getting water resource management right. This book takes us further on that journey.

Professor Geoffrey D. Gooch, PhD

Director, Dundee Centre for Water Law, Policy and Science
(under the auspices of UNESCO)
Professor of Water and Environmental Policy
University of Dundee
DD1 4HN Dundee, Scotland, UK
Email: g.d.gooch@dundee.ac.uk
Web page: http://www.dundee.ac.uk/water/

February 2014

1 Introduction

Purpose of the book

The purpose of this book, *Integrated Water Resource Planning: Achieving Sustainable Outcomes*, is to provide practical guidance on water resource planning through each step of the water planning process, based on the authors' research and experience over several years.

With increasing demands and competing uses for water globally, critical decisions made about water management impact the environment and community health and well-being. Evidence from effects of droughts and floods as well as from heartfelt conflicts about sharing water suggests that we still have a lot to learn about how to achieve sustainable and adaptive management of rivers and groundwater.

In recent years many countries have reformed water institutions. Jurisdictions around the world have been so busy implementing new approaches there has been little opportunity for reflection and sharing of ideas. Early motivation for us to write this book derived from our own experiences working on water reform in different State governments in Australia, with little awareness of what was working well elsewhere. We knew what others were trying but would it work any better than what we were doing?

Furthermore, reviews in Australia have identified gaps in skills and capability that exist or are emerging in the water industries (ICEWaRM 2005). In 2008 we, along with colleague Vanessa O'Keefe, undertook interviews of water planners and stakeholders around Australia as part of the National Water Commission (NWC) funded 'Water Planning Practices and Lessons Learned' project (Hamstead *et al.* 2008a). Water planning agencies in Australia, which have been implementing reforms since 1996, had a high staff turnover and deficit in certain skill areas such as hydrology, community engagement, social assessment, scenario planning and policy. Similarly a survey of water planners in 2009 (Mackenzie and Bodsworth 2009) identified interest in further training and professional development on collaborative water planning approaches, particularly in the areas of Indigenous and cross-cultural engagement, conflict management and social assessment. An accompanying proposed training program identified nine possible deliverers within Australia of a post-graduate/professional course focusing on these social aspects. Further support for expanding tertiary degree curriculum to ensure water professionals are 'work ready', was identified in a 2008 review of universities and the water industry (Murray and Seddon 2008).

Since then the NWC has engaged us and others to contribute to their periodic assessments of progress in implementing Australia's National Water Initiative[1]. These assessments confirmed challenges arising consistently in all jurisdictions, some innovative approaches being tried, but an overall lack of guidance for water planners trying to implement a significant reform in managing freshwater and groundwater resources. The NWC has gone a long way in filling knowledge gaps on specific topics in Australia.

While these reviews demonstrate a need for targeted practical guidance in Australia, our further research suggests that the demand is a global phenomenon. A tendency in countries around the world is a continuing reorganisation of institutional arrangements in tandem with electoral cycles, with some jurisdictions increasing interdisciplinary teams and community engagement and others taking a pragmatic approach to getting an outcome with minimal external input. Continuous organisational change leads to high staff turnover and compounds skill shortages and loss of experiential learning. In many cases, this has disrupted effective monitoring and evaluation of processes and outcomes, which is absolutely essential in assessing program effectiveness. At the same time, assessments of reform and implementation of integrated water resource management (IWRM) globally (UNEP 2012), of the EU Water Framework Directive (Bourblanc et al 2012; EC staff 2012), South Africa's integrated water management approach, and Australia's water reform policy – the National Water Initiative (Hamstead et al 2008a, b), have shown much progress, but reinforce the fact that water professionals and decision-makers around the world are on a steep learning curve.

An international examination of experience of water reforms can provide lessons as jurisdictions plan in a time of heightened awareness of ecosystem needs, climate change, and increasing and conflicting demands on water resources. A greater multidisciplinary and multi-sectoral approach is needed to proactively manage uncertainty of climate, long-term implications of management actions, and the sheer increase in demand from urban growth, irrigation and mining. Just as important, water professionals need to develop better ways of using evidence gained from global experiences, to convince both decision-makers and the community to make hard decisions, weighing immediate objectives with longer-term outcomes. The insights gained from this book should be transposable and instructive for water professionals engaging in water allocation processes worldwide.

---

1 The NWC in Australia has contributed substantially to addressing knowledge deficits in relation to particular themes. While we refer to many of their reports, we highly recommend an avid reader search www.nwc.gov.au.

Our approach

Many recent books on water management focus on specific themes such as climate change or environmental assessment; are location-specific (e.g. Australia or EU based); or are edited texts consisting of contributed articles. Various agencies have prepared guidelines or manuals (DEFRA 2006) for similar processes but in many cases they are not sufficiently detailed or topic-relevant to provide guidance for water planning. This book aims to fill a gap by providing water planners and policy analysts with detailed, practical guidance on water resource planning, supported by the evidence they need to convince others of ways to address the problem. Importantly this book will fill in the gaps for professionals in interdisciplinary teams including sociologists, hydrologists, engineers, ecologists, and community consultation or conflict specialists, by providing a basic grounding in areas outside their usual expertise, including environmental and social impact assessment, consultation, and risk-based approaches.

Many community stakeholders have considerable experience in the water business and act as representatives of natural resource management, conservation, industry (farmers, irrigators) groups, or local government bodies. This book will provide ammunition in their quest to ensure that water planning outcomes are justified and justifiable. Thus it can be used as a reference, academic resource, or in short training or university courses in water resource planning. The book should be useful in advanced courses in natural resource management, regional and infrastructure planning, and engineering.

This book is distinctive in applying IWRM through the lens of a water planning framework based on the elements of the planning cycle, to take a systematic approach to the entire water planning process. We include case studies to improve understanding of the context, practical tools and implementation techniques for achieving sustainable outcomes. Regulatory mechanisms, scenario development based on data, risk assessment, community engagement, and social assessment are discussed.

While to a large extent these practices have broader application, we focus on planning for allocation (or sharing) of water resources, with the aim of achieving sustainable use of water resources, particularly in cases where there is competition for the available water, or 'wicked problems' in water allocation.

Water planning is greatly affected by the legal, institutional and political context in which it occurs. Much has been written on preferred prerequisites for water resource planning. The UN World Water Development Report 3 (UNEP 2009) states that important conditions for implementing IWRM include political will and commitment; well-defined, flexible and enforceable legal frameworks and regulation; adequate investment, financial stability and sustainable cost recovery; and participation and coordination mechanisms. While we touch on these matters, our focus is on the process for water resource planning, recognising at the same time that the means and methods may be constrained by this context.

Our systematic model of a water resource planning process represents a possible 'ideal' but most water resource planning does not happen in such an orderly way, or as thoroughly as would be considered best practice. Priorities change due to crises; compromises are made due to financial constraints; time pressures may lead to skipping or amalgamating steps; and decisions have to be made despite limited information. Our work in preparing this book has highlighted a key issue – few water resource plans and processes have been critically evaluated against their objectives to determine what has been most effective. We wholeheartedly hope that this book will facilitate jurisdictions in the ever so important critical analysis of effectiveness of water resource planning.

Water resource planning as part of IWRM

A term that is commonly used in water management circles is Integrated Water Resource Management (IWRM). In 2002, at the Johannesburg World Summit on Sustainable Development (WSSD), The Technical Advisory Committee of the Global Water Partnership defined IWRM as

> a process, which promotes the coordinated development and management of water, land and related resources in order to maximize the resultant economic and social welfare in an equitable manner without compromising the sustainability of vital ecosystems (GWP TAC 2000: 22).

The WSSD's *Plan of Implementation* included IWRM as one of the key components for achieving sustainable development. It provided specific targets and guidelines for implementing IWRM worldwide (Rahaman and Varis 2005). The Global Water Partnership suggested three goals of IWRM are:

- economic efficiency – to make scarce water resources go as far as possible and to allocate water strategically to different economic sectors and uses
- social equity – to ensure equitable access to water and to the benefits from water use, between women and men, rich people and poor, across different social and economic groups both within and across countries
- environmental sustainability – to protect the water resources base and related aquatic ecosystems and more broadly to help address global environmental issues (Jønch-Clausen and Fugl 2001).

IWRM recognises that water resources have multiple uses that interact, thus requiring holistic consideration of all uses. These uses include human consumptive purposes as well as the provision of water for biodiversity and the preservation of ecosystems. IWRM recognises that water behaves within a complex hydrologic cycle that includes rainfall and evapo-transpiration; runoff and infiltration; and surface and underground flow. Thus in managing one aspect of the cycle, consideration must be given to the interaction with other connected aspects, for example between connected rivers, between rivers and connected aquifers, and between runoff affecting forestry activities, vegetated landscapes and downstream rivers. It recognises that terrestrial and underground water flows do not take any notice of national or state boundaries, so management in one jurisdiction interacts with that in others that share the same water system. Further, IWRM acknowledges that management aimed at other purposes, for example development for cities, agriculture, electricity, and tourism, can affect management of water resources. IWRM does not however insist on managing all of these together in one process, as this would become intractably large and complex. Rather it aims for management to occur at a scale and scope that is practical, but at the same time providing bridges and connections to management processes where interaction is likely (Lenton and Muller 2009).

For our purposes, water resource planning focuses on how water is allocated for different purposes, as an aspect of water resource management. In this book we assume that it shares the same aim as IWRM – to achieve a combination of economic, social, and environmental objectives. As much as possible all are achieved together, but in reality there is normally a significant degree of competition between these objectives, so 'optimal' or 'balanced' solutions are sought that reflect a realistic trade off based on human values. Water resource planning aims to reflect the integration aspect of IWRM, either directly managing or providing connections to activities that interact.

The growing global population will increase competition for domestic, agricultural and industrial water consumption leading to stress on water resources. Solutions must be locally based since each country experiences its own issues with development needs, water stress, and at different times: floods or drought, lack of potable water, or degraded resources. These shape the way a water resource planning process should be run. Doing nothing is a high-risk option. Planning is essential.

The scale of water resource planning can vary. While it should aim to maximise opportunities for integrated management, the decision has to be based on what is practical and achievable. Internationally it is increasingly common to place an emphasis on planning at the river basin scale because of the strong interaction between upstream and downstream uses of water. However scope varies substantially on whether linked aquifers are included; whether planning extends to activities in the catchment (e.g. forestry, land clearing) affected by water; whether it extends to urban supply (e.g. demand management, sanitation, pollution, and use of recycled water); and so on. Where such matters are not included directly in the same planning and management system, they are often addressed through linkage arrangements such as overarching strategies, requirements to consider effects outside the immediate scope of the plan, or compensatory packages to enable trade-offs.

Each jurisdiction has its own way of addressing water resource planning. In Australia, planning for water allocation is governed by its own process and legislation, separate from water pollution, with both linked in varying degrees with 'catchment or natural resources planning', depending on the state. In the European Union, River Basin Management Planning encompasses planning for all of these purposes, although legislation and processes for doing so vary with each member.

Challenges for water resource planning

To address the goals of IWRM – economic efficiency, social equity and environmental sustainability – our analysis of the literature suggests common challenges that need to and can be addressed through water resource planning are:

- better use of existing resources
- ecosystem and environmental quality
- uncertainty about the future and implications for water security
- conflict due to inequitable distribution of costs and benefits about water allocation.

Better use of existing water resources

The highest priority for use of water all over the world is for domestic use, for drinking and sanitation. This is reinforced by Millennium Development Goals to improve universal access to drinking water; more specifically to reduce by half the proportion of the population who do not have access to clean drinking water by 2015, while ensuring that the environment is conserved (UNDP, 2000). This priority for domestic use is also reflected in the trade-offs frequently made in developed countries which direct water to urban use during water shortages.

The OECD Environmental Outlook projects that by 2030, 3.9 billion people, or 1 billion more people than today, will be living in areas under severe water stress, unless present water policies are improved. The majority of these people will be in emerging economies, particularly in South Asia and China. Water stress reflects an imbalance between growing and competing water demands, and increasingly erratic availability of freshwater of sufficient quality (Gurría 2009).

It is generally agreed that resource scarcity and technology are not the only causes of this water crisis, but past and current inefficient management in nearly all countries of the world are also responsible (Pahl-Wastl et al. 2008; Biswas 2009; Gurría 2009). The poor service reliability of good quality reticulated water in many developing countries is attributable to poor governance (Biswas 2009). Phnom Penh, in Cambodia, for example, managed to achieve drinkable water supply on a 24-hour uninterrupted basis within the space of four years as a result of change in governance involving leadership and political non-interference (Biswas and Tortajada 2010). Improved service provision and better governance directly affect the financing gap: increased customer satisfaction improves willingness to pay for the service as well as financial institutions' willingness to provide loans (Gurría 2009).

Effects of drought world-wide are severely under-recorded in spite of the clear impacts on agricultural production, rural livelihoods, food security, mortality, migration, conflict and ecosystem decline. It is claimed that drought ranks as the single most common cause of severe food shortages in developing countries (Below *et al.* 2007).

This has implications for managing drought impacts through water resource planning by considering water use efficiency, water storages, and climate forecasting. For nearly all agricultural products, water requirements for producing a unit of any product is 50–200 per cent more for non-efficient farms as compared to their more efficient counterparts (Biswas 2009: 404). Targeted interventions are necessary to reduce the risks associated with rain-fed agriculture, especially those in marginal rainfall areas often managed by poor smallholder farmers (Mati 2007). Man-made reservoirs play an important role particularly where precipitation is seasonal or erratic, enabling the storing of water for use in drier periods. Increasing the number of water storages thus can reduce vulnerability due to climatic variability, especially water risks associated with both floods and droughts (Mati 2007). Regions with the highest likelihood of needing an increase in reservoir storages are Asia and Sub-Sahara Africa, the latter for agriculture, domestic and industrial needs (White 2010: 55). Storages can be a solution in some areas, however the amount of land taken by storages and diminishing availability of good sites can be a constraint, along with losses due to evapo-transpiration and sedimentation over the short and long-term.

Box 1.1: Competing demands: Tana River, Kenya

The Tana River runs from Mt Kenya through national park and small agricultural holdings, supplies Nairobi's water supply and hydroelectricity through 4 dams, then provides water to small communities and cattle in a semi-arid area to the coast near Lamu. With an emphasis on Millennium Development Goals to provide potable water, the government is encouraging water harvesting and development (another large dam as well as local ones). Figure 1.1 (see colour plates) for example, illustrates a small new 'sand dam' for local use in a rural community in Kenya, with questionable longevity as a result of sedimentation due to adjacent land use. A water planning process run by government in conjunction with community-based organisations

(water-user associations) needs to consider environmental flow requirements to ensure sufficient water for communities at the bottom of the
catchment. The Lamu water supply is from mining of dunal aquifers
which are at risk of salt water intrusion.

While the economic advantages of dams for irrigation, water supply, hydroelectricity, flood control, navigation, and fish farming are acknowledged, the social
and environmental impacts of large dams need to be better addressed (White
2010). Furthermore the benefits of large dams in particular need to be realisti-
cally evaluated to make a fair trade-off with impacts: many hydro-power dams
underperform compared to power generation predictions providing smaller
revenues for paying off debts incurred to build the project (McDonald 2009).

Pressure for increasing water supply often comes from the need to increase
food production. In terms of water for feeding the world's population, it has
been argued that accessing land and water resources as well as fertilisers to
increase production is less economic than achieving better distribution and
less loss of available food. While India produces 15 per cent of the world's
fruit and vegetables much of the resource is lost due to poor storage, transportation and supply chain issues. Only 2 per cent of this produce is processed,
compared to 70 per cent by Brazil and 80 per cent by Malaysia (Biswas
2009). In developing countries 40 per cent of losses occur at post-harvest and
processing levels while in industrialised countries more than 40 per cent of
losses happen at retail and consumer levels. Food losses during harvest and in
storage translate into lost income for small farmers, so reducing losses could
therefore have an 'immediate and significant' impact on their livelihoods
and food security (Cederberg *et al.* 2011). Thus, proper storage, distribution
and management of the food that is already being produced should be a key
priority for highly populated countries such as India and China, if increasing
the availability of food to consumers is the main objective.

The recent changes in practice by industry (the next major global user
of water quantitatively), driven by pricing, illustrate that there is enough
knowledge, experience, technology and even funds to solve many of the
world's water problems. For example, Figure 1.2 (see colour plates) illustrates
a more efficient form of rice growing, 'Sustainable Rice Intensification' (SRI),
that uses less water and other inputs to produce greater yields. Figure 1.3
(see colour plates) shows water-efficient irrigation that minimises water loss
from evaporation. Even the most water-efficient steel plant uses only 4 per
cent of the water of the least efficient plants by using improved designs, good
technology and better management practices, including extensive recycling
of water (Biswas 2009: 404). Furthermore, water as a renewable resource can
be used and reused many times, so managing it well is as important as how
much freshwater we have. Effective use of stormwater and water harvesting

from roofs for non-potable uses (e.g. Figure 1.4 in colour plates), and more cost and energy-efficient technologies for recycled and desalinised water will assist in meeting urban and industrial needs. Just to maintain services at current levels, though, many developed countries estimate an increase of between 20 to 40 per cent in water spending as a share of GDP (Gurría 2009).

The need to provide potable water and food for the increasing global population puts tremendous pressures on governments, communities, and individuals to develop water resources. The opportunities for storages in some parts of the world are underdeveloped, and in all cases, the impact on long-term sustainability of watercourses needs to be addressed. The purpose of a storage will have an effect on flows: off-stream storages for irrigation result in a direct reduction in flow; whereas flows through a hydroelectric scheme need to be carefully managed to maximise generation while achieving as close to a natural environmental flow as possible. So a key message is that good governance and clever on-site water management is integral to making the best use of existing water resources. Any strategies for sharing water need to include (1) examination of efficient use and distribution; (2) adequate supply, demand and environmental data; and (3) community involvement and education.

Maintaining ecosystems and other public benefits

Just as water is essential for human life; it also sustains ecosystems which maintain biodiversity and provide many other public benefits. The increasing extraction of water and regulation of rivers over the years has contributed to substantial net gains in human well-being and economic development, but has also resulted in drying rivers and streams, falling water tables, and shrinking lakes. In fact, changes in ecosystems have a more direct influence on human well-being among poor populations than among wealthy populations (Corvalan et al 2005: 48).

The gains have been achieved at growing costs in the form of:

- the degradation of many ecosystem services
- increased risks of nonlinear changes
- the exacerbation of poverty for some groups of people, often those lower in a river catchment.

Unless addressed, these problems will substantially diminish the benefits that future generations obtain from ecosystems and are a barrier to reducing global poverty and achieving the Millennium Development Goals. Water resources are not infinite and their use comes with an environmental price tag: over-exploitation (depletion of supply) and degradation (deterioration in quality) are two aspects of the price to pay (UNEP 2009). The cost of reversing the degradation of ecosystems is much higher than preventing loss and damage in the first place but even restoration provides economic benefits worldwide (Menz *et al.* 2013). Provision of ecosystem services for human welfare is dependent on sustainable and properly functioning ecosystems which in turn provide water security over the long term (UNEP 2009).

Figure 1.5 (see colour plates) illustrates a creek bank protected by riparian vegetation from the intense rainy season flows and power boat wash. Without such protection, fish habitat and recreational fishing and boating would be diminished over time by silted up creeks.

Most OECD countries have addressed surface water pollution, mainly by regulating industrial discharges and investing in urban sewage treatment. The 'polluter pays' principle and source control have increasingly gained support over technical end-of pipe solutions (Pahl-Wostl *et al.* 2008: 2). But, according to Gurría (2009), diffuse pollution from agriculture (e.g. fertilisers, pesticides and livestock manure) still remains a challenge. Nearly 60 million tonnes of nitrogen are expected to reach coastal waters from inland sources by 2030, killing fish and disrupting ecosystems along the way. There is growing awareness that the need for integrated approaches requires taking a range of trade-offs into account and involving stakeholders in the whole management process (Pahl-Wostl *et al.* 2008: 484).

Box 1.2: Ecosystem services offer benefits beyond the environment

Ecosystem services can be delineated in four ways:

- Provisioning services – the products from ecosystems such as food, fresh water, timber and fuel, fibres and genetic resources;
- Cultural – recreation, transport, ecotourism, spiritual, religious and aesthetic uses, education, cultural heritage and a sense of place;
- Regulation – climate regulation, flood alleviation, water purification, and disease regulation;
- Supporting – underlie the above, and include nutrient cycling, soil formation and primary production (UNEP 2009: 10).

Major management options to address ecosystem functioning and services include:

- Maintaining environmental flows;
- Pollution control;
- Ecohydrology and phytoremediation;
- Habitat rehabilitation;
- Conjunctive use of surface and groundwater;
- Watershed management;
- Water demand management; and more recently
- Payment for ecosystem goods and services (UNEP 2009: 18).

Box 1.3: Managing for variable flows and water quality

The Mara River is a typical example of the inter-relationship between consumptive use, ecosystem services, and human benefits. The Mara River runs from Mt Kilimanjaro through a coffee-growing area, Masai Mara Nature Reserve, an agricultural area in Tanzania and ends up in Lake Victoria, which flows into the upper Nile basin. Environmental flow assessments have been completed for both the Kenyan and Tanzanian parts of the river, but are somewhat limited by available data. The challenge for water resource planning is to balance water use for native animals, which are a major tourist attraction and income generator, and for human use including potable water, irrigation and cattle grazing. While maintaining a minimum water flow is a typical objective in most water plans, in Masai Mara Nature Reserve hippopotamuses (a tourist attraction) need 1.5m depth of water in pools for their health (Figure 1.6 in colour plates). Nutrient levels in the river run high from: wildebeest deaths while crossing the river during migration, wild and domestic animal use, agricultural run-off, and tourism facilities (although waste treatment for the latter are licensed). The Mara River, along with other streams, contributes to the nutrient load of Lake Victoria which has massive growth of the invasive water hyacinth.

A major driver for Australia's water reform was the 1991 toxic algal bloom in over 1000 km of rivers in the Murray-Darling Basin, which affected stock, domestic, tourism and recreational use of water (Figure 1.7 in colour plates). Further motivation for addressing the land–water nexus was identification of a potential three-fold increase in the area affected by dryland salinity from the current 5.7 million hectares (NLWRA 2000). Salinity had already had impacts on water quality, native vegetation, and lifespan of road, rail and urban infrastructure. Similarly, a decline in the Great Barrier Reef off the coast of Australia has been partly attributed to nutrient and sediment run-off from the adjacent mainland. This has resulted in a $AUD200m collaborative Reef Rescue Program with farmers and graziers to reduce diffuse contaminants into the Reef World Heritage Area.

Degradation has also prompted reform in other developed countries. In 2005, the province of Alberta in Canada announced it would no longer accept applications for new allocations from the Bow, Oldman, and South Saskatchewan River sub-basins because of river degradation. At that time, 22 of the 33 main Alberta rivers were suffering moderate environmental effects from increased water stress caused by water extraction for consumptive use, five were suffering heavier environmental effects, and three were environmentally degraded. Of note is that about 75 per cent of all allocated water is used for irrigation, a large percentage of which is used to produce relatively low-value crops (Bjornlund 2010). As a consequence, the extraction of surface water and hydrologically connected groundwater for all forms of consumptive use was capped.

Even eastern Canada, which has traditionally been seen as having a plentiful water supply, has had increased problems. Recent mid-summer base flows in the Grand River, Ontario, which supplies several mid-sized cities, were primarily comprised of treated sewage. Pollution of a municipal groundwater source from diffused agricultural sources caused death and illness in the town of Walkerton in 2000 and prompted a considerable response from policy-makers (Ferreyra *et al.* 2009).

Box 1.4: Managing urban water supply and pollution in southern Ontario, Canada

Grand River, Ontario

The Grand River is characterised by seasonal flow of heavy spring runoff with a low end-of-summer flow primarily maintained by treated sewage discharged by the many small towns and cities in southern Ontario that it runs through on its way to Lake St Clair, the smallest of the Great Lakes. It is Canada's only designated 'Heritage River' in a built-up area, designated partly due to historic mills, a canal, and other man-made waterway features. The management body, the Grand River Conservation Authority (GRCA), is trying to restore original flow patterns to some extent, which have implications for water use restrictions and historic mill races. The GRCA deals with multiple well-informed stakeholders and relies on provincial government agencies and local community groups to implement many of its actions, although it is unique in having its own source of funding from small hydropower projects and camping fees.

Walkerton, Ontario

The case of Walkerton is an example of a government policy and legislative response to an environmental disaster that killed 7 people and made 2300 seriously ill in southern Ontario in May of 2000. One of the wells supplying the town of Walkerton (pop. 5000) became contaminated with *E.coli* as a result of a nearby feedlot effluent draining into the aquifer recharge area, known for years as a potential source of contamination. An inquiry, known as the Walkerton Commission, found the Walkerton Public Utilities Commission operators to be grossly negligent. The preliminary direct economic impact was estimated at $65M. Recommendations focused on source water protection and a comprehensive multi-barrier approach, among other things.

Floods, droughts and contamination affect all countries of the world. Kenya has an approach to protect its 'water towers', upland sources of water. Collaborative approaches to address water resources are also in place in the European Union, through the Water Framework Directive. This provides a consistent approach to water management across member countries through River Basin Management Plans and supporting programs such as NeWater, which researched and fostered approaches to adaptive water management. The trend is to 'whole of catchment' approaches to water management. Changing community values (as evidenced by recent protests in China) are leading to increasing pressure to avoid pollution and maintain instream flows for environmental, recreational, and cultural purposes for health and well-being.

A key message is that strategies for water resource plans need to be developed within a whole-of-catchment context, considering land use and forest cover which affect the retention of water upstream to reduce flooding and cost of cleaning contaminants and sediment for downstream use. Continued health of rivers need to address in-stream needs, environmental events, and minimum flows before allocation of water for consumption. These all affect the economics of water use for purification, flood mitigation and recovery, and other public benefits such as subsistence and commercial use of fish resources and recreational use. It has implications for the type of people – stakeholders – that need to be involved in a water resource planning process so that local knowledge and community values are included. It also mandates decisions to be based on sound information.

Uncertainty about the future and water security

Appropriate investment in water infrastructure, either on a regional level for water supply and flood prevention or on-farm, is dependent on an understanding of river and groundwater characteristics and factors that affect them, such as land use and climate variability.

Future water problems will be different from the past. Climate change is expected to result in increased frequency of extreme events, with both higher rainfall and drought in different locations, creating uncertainty in being able to reliably predict water availability. Significant reduction in river runoff and aquifer recharge is expected in the Mediterranean Basin and in the semi-arid areas of the Americas, Australia and Southern Africa, affecting water availability and quality in already stressed regions due to reduced precipitation and high evapo-transpiration (FAO 2012, White 2010). Effects can be intensified by consumptive demands and alteration in flow due to extraction or other modifications. Figure 1.9 (see colour plates) illustrates the result of a drought exacerbated by reduced flow from upstream extraction in a semi-arid area reliant on overbank flooding from intense rainfall higher in the catchment.

OECD countries have already experienced significant flood damage and droughts often within the same region in different years. Examples include both drought and flooding in Queensland Australia, parts of the United

States, and southern Europe. Flooding is regularly experienced in the Grand River Canada, the southern Scottish town of Peebles (Figures 1.10 and 1.11 in colour plates), and Tisza river basin in Hungary, among many others. Flooding impacts range from inconvenience to destruction of property and infrastructure at great cost to individuals and governments, through to deaths in highly populated areas such as Bangladesh, and even Beijing in July 2012. Despite some progress, neither OECD or non-OECD countries have fully integrated climate adaptation into domestic water policy frameworks (Gurría 2009).

The relationship between water and other sectors affects levels of uncertainty and water security. Land uses including vegetation clearing affect quality and quantity of run-off, but time frame and severity of impacts are difficult to predict. Likewise the impact of changes in biodiversity on ecosystem services, and then on human well-being are difficult to quantify and predict. Using biofuels to deliver 5–6 per cent of total energy production would double water withdrawals for agriculture (Gurría 2009). This may be greater than can be sustained and still deliver food security and potable water. Likewise calculations about construction parameters, capacity, and safety of water storages are all affected by the limitations of hydrological models (Fiedler and Doll 2007).

Risk and vulnerability need to be taken into account in any water resource planning process, as well as exploration of possible impacts. Typical of natural resource management challenges are uncertainties around quantification and relationship between cause and effect. Exploring a range of future scenarios can help avoid surprises and provide consideration of contingencies. Ongoing adaptive management is needed, requiring effective monitoring and evaluation of the effectiveness and efficiency of management practices as well as the changing situation.

Conflict due to inequitable distribution of costs and benefits

A key driver for a fair process for water resource planning is conflict or potential for conflict. This may occur due to high competition for increasingly scarce water resources. Conflict is often generated through perceptions of lack of procedural fairness and distributional equity. In some cases, physical and economic water scarcity around benefits accruing to some at the expense of others, create a venue for winners and losers, and inequitable distribution of costs and benefits. This can easily be the case due to the physical location on a watershed upstream or downstream. It can also occur in cross-boundary watersheds, where jurisdictions have a prime responsibility to service their own population, at the expense of a neighbouring one.

Many cases have been documented where people or groups of influence or areas of high priority to government have gained a distributional advantage, due to the goal of promoting economic development or a sometimes poorly-defined 'public benefit'. The legitimacy of such decisions is often supported by governments for a number of reasons:

- to promote rural and regional economic development through mining or mono-culture for food, bio-fuel, and flex crops, especially in marginal agricultural areas;
- to meet a food or water shortage;
- to improve hydropower, water supply to urban communities, commercial businesses or balance of trade.

Hydrologic complexity and surface water/groundwater interactions often make it difficult to prove the associated negative impacts on the environment and diverse social groups (Sosa and Zwarteveen 2012). Meanwhile the benefits are often over-exaggerated: created jobs might last less than year and fail to build local skills, a water supply dam may be the least desirable of other options, or a hydroelectric dam creates lower than expected output of electricity generation. Examples abound, including dams in Laos PDR (Matthews 2012), mining in Cajamarca Peru (Sosa and Zwarteveen 2012) and coal seam gas development in areas around the world. Such situations are not unique to developing countries, although poorer people and those living within authoritarian regimes have less ability to influence a better outcome. In some cases because of poor governmental institutional capacity, private companies also become responsible for water allocation and safeguarding water quality, which raises potential transparency and accountability issues and conflict of interest (ibid).

During water resource planning processes, conflicts have to be managed, but many involved in the water sector have little experience with managing conflict without use of power. Poor governance practices create opportunities for what is known as 'water grabbing' 'where powerful actors are able to take control of, or reallocate to their own benefit, water resources already used by local communities or feeding aquatic ecosystems on which their livelihoods are based' (Mehta *et al.* 2012: 197). Opportunities are created for influential actors, while local livelihoods and the environment, are negatively impacted without compensation.

Good governance practice can ensure inclusion of marginalised, disadvantaged and Indigenous groups in decision-making. Information needs to be provided in user-friendly ways tailored to the local environment to foster informed decision-making (Figure 1.12 in colour plates). The range of social, economic as well as environmental impacts of a range of options also needs to be considered. Good facilitation, participation, consensus building and if necessary, conflict resolution procedures, will ensure all parties are fairly heard. Exploring options for trade-offs should be a requisite part of the process.

A key message is that critical decisions about water allocation should be made within the context of a catchment water resource plan that considers sound information on social and economic benefits and impacts, aims to minimise negative impacts, and actively seeks and listens to the voices of all persons, not just powerful lobby groups or commercial operations. The skills of interdisciplinary teams should include partici-pation and conflict management.

Book structure

In the remainder of this book we set out principles and practices for devel-oping water resource plans to address these challenges and better achieve the three goals of IWRM – economic efficiency, social equity and environmental sustainability.

In Chapter 2 we review international and national principles and policies and show how they influence the development of water resource plans.

In Chapter 3 we discuss conceptual frameworks for describing the planning process and the internal logic of plans. These frameworks are used in the rest of the book as a structure for discussing different aspects of water resource planning. Because there is no consistent terminology for these things, we describe the terms we use in this book.

Chapter 4 addresses community consultation and collaboration. We explain why it is important for water resource planning, how it can be done at different steps of the planning process, and different methods that can be used in different circumstances.

Chapters 5 to 9 expand on the water planning process described in Chapter 4, setting out in detail methods and approaches. Chapter 5 covers situational analysis, where the current knowledge of the water resource and its uses is gathered, and risks and opportunities are identified. In chapter 6 we explain how objectives and logic can be determined, providing the framework for further developing the plan. In Chapters 7 and 8 approaches to identifying, comparing and selecting actions to achieve the plan objectives are described. Chapter 9 then discusses how monitoring and evaluation arrangements can be established to provide for ongoing adaptive learning and improvement.

Lastly, in Chapter 10 we make some concluding observations.

2   Guiding principles

Global principles for water management have emerged through international discussions and negotiations about common internationally recognised challenges to achieving the three sustainability oriented goals of IWRM – economic efficiency, social equity and environmental sustainability – outlined in chapter 1. They are refined in treaties at international forums such as the Rio conference in 1992 and are adopted in agreements, legislation, policies, and institutional arrangements such as Agenda 21 and the EU Water Framework Directive. They are often reflected in bilateral agreements between countries that share watercourses and are endorsed at country level through legislation and policy. The important point though is that, while small steps can be taken in an ad-hoc manner and without a legislative directive, a consistent broadly applied policy and legislative framework provides a mandate and justification for applying the human and financial resources to achieve water sustainability goals.

This chapter reviews key principles that have informed global policies and conventions and are implemented by individual countries in nationwide agreements, such as Australia's National Water Initiative, and through national legislation such as South Africa's *National Water Act 1998*. We illustrate how core principles influence sub-national/state/province level legislation and programs, using Australia's Murray Darling Basin as a case study. We also highlight fundamental issues that should be addressed in transboundary agreements, whether between countries, or within countries across states. We suggest that a sound water resource planning process can provide a mechanism through which countries/states and key stakeholders can work together to find solutions and prevent or resolve disputes.

Principles that inform policy

For each of the major challenges described in the previous chapter, principles can be developed to guide decision-making. Principles are often interrelated: ignoring one may affect others. Principles of global significance are encompassed in international treaties or conventions. While these agreements often relate to cross-boundary watercourses, they also acknowledge the fact that even if watersheds are not interconnected, how water is dealt with in one country affects others. For example, poor water management that results in food shortages creates pressure to divert aid, engenders conflict or influences migration.

Principles can be expressed as formal or informal rules reflecting a system's purpose. They can be enshrined in written law as a legal obligation, to some extent limiting individual freedom, ostensibly for the greater good. Alternatively principles that reflect society's values and norms may be implemented through normative behaviour, peer pressure, and education. In between these two levels of governance are regulations which give effect to law; and policies, strategies, detailed guidance, or codes of conduct formed by government, river or catchment management bodies, industry bodies or NGOs which advise on priorities, behaviour change and implementation processes.

Each of the key challenges to be addressed through water resource planning described in Chapter 1, can be expressed as principles.

The first principle is to *make better use of existing resources*. To embrace water use efficiency we need to understand:

- the characteristics of the water resource, available water, and how to estimate reliability and security;
- how much water is used for various purposes, and how it is distributed and used (e.g. flood irrigation);
- latent or unmet demand;
- whether demand management and water use efficiency mechanisms are in place and what mechanisms would work to reduce existing use and demand such as pricing and competitive water markets;
- how water users can be engaged to supplement existing data with local knowledge, consider efficient water use mechanisms and possible incentives, and support monitoring of effectiveness.

The second principle, to *achieve ecosystem and environmental quality*, includes consideration of:

- water management within a whole-of-catchment or watershed basis;
- land use affecting flow and quality of watercourses and groundwater;
- sound data on in-stream ecological needs, interconnected surface and groundwater systems, creating or allowing environmental events and minimum flows to maintain the environmental asset;
- trade-offs which take account of short and long-term implications for social, environmental and economic costs and public benefits (e.g. cost of sedimentation for water purification; flow effects on fish for subsistence, commercial, or recreational use);
- how relevant stakeholders can contribute local knowledge and community values to the decision-making process

- a regulatory process (such as licensing of water use) that reserves minimum flows to allow for natural environmental events before extraction, and a defined consumptive portion that maximises the economic benefit and minimises the socioeconomic impact.

The third principle, to *address uncertainty about the future and aim for water security*, suggests a need to understand:

- resource characteristics and risk and frequency of environmental extreme events
- vulnerability of property and humans to existing and potential threats
- a wide range of potential future scenarios and management options
- consideration of mechanisms that clarify level of security (e.g. licenses)
- reflective monitoring and evaluation based on a suitable logic frame that allows for adaptive management
- how to engage and communicate with communities about uncertainties.

The fourth principle, to *reduce likelihood of conflict due to inequitable distribution of costs and benefits about water allocation*, requires:

- fair participatory processes and skilled facilitation that ensure all stakeholders are appropriately informed and have a voice, including marginalised, disadvantaged and Indigenous groups;
- transparent and accountable processes for considering options and trade-offs based on sound environmental, economic and social information;
- mechanisms to equitably allocate benefits derived from water, not the water itself (i.e. 'benefit sharing') and to reduce negative impacts.

These four principles overlap and complement each other. By aiming to achieve these principles through a water resource planning process, we can address the range of challenges and better achieve the three goals of IWRM.

Water resource planning occurs within the context of international, national and state policies that guide the planning process and the manner in which environmental, social and economic outcomes are balanced or traded-off in the decision-making process. While water resource planning should lead to sustainable management and use of water, there is considerable debate about what this means in a practical sense.

How would you embed these principles in institutional mechanisms, to ensure implementation? In the remainder of this chapter, we examine some influential examples of international and national frameworks governing water resource management that encompass these principles. We end with a set of guidelines for effective water resource planning, derived from these international and national lessons. The chapters that follow provide insight, concrete guidance and examples of how to implement the principles and guidelines through a water resource planning process.

International conventions and agreements

Principles of international law govern relationships between nations and are fundamental to the charter of the UN and maintaining international peace and security (Liguori 2009). Key themes of particular relevance to water resource planning relate to the principle of 'no significant harm' to environment and others, and 'benefit sharing' (Abseno 2009), as well as employing 'good faith', which implies honesty and fairness between parties who are seeking a solution.

International agreements take many forms including conventions and transboundary binding and non-binding agreements. A minimum number of signatories are often required for international treaties and conventions to come into effect. On committing to an agreement, a government uses it as guidance in making relevant laws, policies, strategies and instruments, and in implementing programmes.

Fundamental to international agreements about water is the concept of sustainable development. It has been a quarter of a century since a generally accepted and succinct definition of sustainable development gained prominence, in 1987, through Our Common Future (the Brundtland Report), which defined it as:

> development that meets the needs of the present without compromising the ability of future generations to meet their own needs (World Commission on Environment and Development 1987: 37).

Since then international agreements have aimed at addressing environmental decline resulting from rapid economic development, by considering the long-term environmental effects, and the consequences to future generations. Water was the focus of a conference in Dublin held in the lead-up to the Earth Summit in Rio in 1992. Water resource management is addressed in chapter 18 of Agenda 21 (UN 1992), an outcome of the Rio conference. It includes a non-binding action plan for improving the state of the world's natural resources with a goal to ensure that supply and quality of water are sufficient to meet both human and ecological needs worldwide. The 2002 World Summit on Sustainable Development called on all countries to establish 'plans for integrated water resources management and water efficiency by 2005'. Soon after, many countries put in place steps to honour this commitment by adopting new national water policies inspired by IWRM principles and basically codifying international water principles (Petit and Baron 2009). The *UN-Water Status Report* (UNEP 2012), assembled in preparation for the Rio+20 conference, notes that 80% of countries have progressed integrated approaches and reforms (e.g. laws and policy) based on Agenda 21. Yet with only 65% of countries having integrated water resource plans and 34% reporting advanced implementation, more work is needed on putting policy into practice (UNEP 2012: 17).

Influential global institutions have also been formed to support these international commitments. The Global Water Partnership (GWP) was founded in 1996 mostly in response to the Rio statement (1992) to promote IWRM based on public participation. Its principles reflect 'equitable and efficient management and sustainable use of water', that water is a public good with social and economic value in all its competing uses, as well as inclusiveness of women. In 2001 it launched a 'toolbox' to provide technical support. Also in 1996, The World Water Council was established as a multi-stakeholder network of water specialists and organisations to build awareness and political commitment to facilitate environmentally sustainable use of water. It convenes the World Water Forum every three years and acts as a think tank for water resource issues.

Table 2.1 summarises key international conventions, their purpose and principles relevant to water management. These express 'umbrella' principles on which there is general agreement among signatories. They have guided nations in developing policies and legislation.

Listed in the table is the unratified 'UN International Watercourses Convention'. Signed by only 18 countries, criticisms have targeted the apparent precedence given to the principle of equitable and reasonable utilisation over the rule of no significant harm; lack of clarity to assist in resolution of upstream–downstream issues; and mandatory dispute resolution being placed in a framework convention (Rieu-Clarke and Loures 2009; Salman 2007). A suggested benefit of it coming into force is that it would support the implementation of other multilateral environmental agreements that touch on water-related issues such the Ramsar Convention, the World Heritage Convention, the Convention on Biological Diversity, and the UN Framework Convention on Climate Change. It could serve as an added political impetus in establishing and, where necessary, strengthening regional, basin and sub-basin agreements (Rieu-Clarke and Loures 2009). However it is noteworthy that the three main contentious elements are common yet critical challenges of water resource planning in general. If an acceptable process for water resource planning can be shown to be effective, it might reassure potential signatories that resolution is possible.

Bilateral or multi-party agreements are also made in regards to transboundary river basins. Globally there are approximately 263 transboundary river basins[1] which involve 145 countries (Rieu-Clarke and Loures 2009). About one-third are shared by more than two countries, and 19 involve five or more sovereign states. For example, the Danube has 17 riparian nations (Giordano and Wolf 2003). Twenty-three countries in Africa have at least three-quarters of their area in transboundary basins and 39 nations have more than 90 per cent of their territory within such basins, including countries such as Zambia and Bangladesh (Wolf *et al.* 1999). When water resources

---

1 Also referred to as internationally shared watersheds.

Table 2.1 International conventions and agreements

Agreement (date)	Purpose	Coverage	Core principles relevant to water
The Convention on Wetlands of International Importance 'Ramsar Convention' (1971)	To halt the worldwide loss of wetlands and to conserve, through wise use and management, those that remain. To encourage designation of sites containing representative, rare or unique wetlands, or wetlands that are important for conserving biological diversity.	164 contracting parties; 2098 wetlands listed (as of 2013)	• Planning and implementing 'wise use' of wetlands • International cooperation • Maintenance of their ecological character, achieved through the implementation of ecosystem approaches, within the context of sustainable development (Resolution IX.1 Annex A (2005)
'Bellagio Draft' Treaty Agreement concerns use of Transboundary Groundwaters – proposed (1989), but *not signed*	To achieve joint, optimum use of the resource, facilitated by procedures for avoidance or resolution of differences over shared groundwaters in the face of the ever-increasing pressures.	Not signed but intended as a model for agreements on management of transboundary groundwaters shared by several countries, e.g. North-eastern African aquifer underlies Libya, Egypt, Chad, and Sudan; a vital European aquifer underlies the Rhine; Israel-Jordan-Syria region	• Use, protect and improve quality and control of groundwaters on an equitable basis • Integrated approach to conjunctive use of surface and groundwater in the border region • Creation and maintenance of an adequate data base • Comprehensive management plans including pumping plans, plus drought management plans • Enforcement within the territory of each party is the responsibility of that party • Biennial review of water quality and quantity control measures • Database of transboundary groundwater resources including quality, quantity, recharge rates, interaction with surface waters • Consideration of declaring transboundary groundwater conservation area • Notification of public health hazards due to contamination • Amicably resolve differences

Dublin Statement on Water and Sustainable Development – the 'Dublin Principles' (1992)	An action agenda was developed based on 4 guiding principles, adopted as a precursor to 'Rio'. Recommended actions were aimed towards: • Alleviation of poverty and disease • Protection against natural disasters (e.g. floods and droughts) • Water conservation and reuse • Sustainable urban development • Agricultural protection and rural water supply • Protecting aquatic ecosystems • Resolving water conflicts Implementation would require investment in the knowledge base and capacity building.	500 participants of 80 organisations attend the conference in Dublin in January 1992	1: Fresh water is a finite and vulnerable resource, essential to sustain life, development and the environment 2: Water development and management should be based on a participatory approach, involving users, planners and policy-makers at all levels 3: Women play a central part in the provision, management and safeguarding of water 4: Water has an economic value in all its competing uses and should be recognised as an economic good
Convention on Biological Diversity (1992) UN Conference on Environment and Development (UNCED), also known as the Rio Summit and Earth Summit in Rio de Janeiro in June 1992	Broad outcomes regarding the environment were reflected in the Rio Declaration on Environment and Development. It has three main goals: the conservation of biological diversity, the sustainable use of its components, and the fair and equitable sharing of the benefits from the use of genetic resources.	193 parties	• Sustainable use • Precautionary principles • Conservation of biological diversity • Prior informed consent of parties providing genetic resources and traditional knowledge • Benefit sharing from research and commercial development of genetic resources • Impact assessment

Agreement (date)	Purpose	Coverage	Core principles relevant to water
Agenda 21 (1992), Rio Earth Summit	International framework agreement and action plan aimed at: • Alleviation of poverty and promoting sustainable settlement; • Conservation of biological diversity, combating deforestation, protecting fragile environments; • Strengthening role of all stakeholders including children, women, Indigenous people, and farms; • Implementation by education, science, institutional and legal structures, data and information and the building of national capacity in relevant disciplines.	Voluntary non-binding; 178 countries adopted; reaffirmed commitment at Rio+20 (2012)	• Sustainable use • Ecosystem protection • Integrated decision-making at all levels • Community involvement
UN Convention on the Law of the non-Navigational Uses of International Watercourses (1997)	To govern and inform the use, management, and protection of the world's international watercourses, for present and future generations, taking into account the special situations and needs of developing countries and affirming the importance of international cooperation and good-neighbourliness. Watercourse refers to system of surface and groundwaters.	Not yet ratified; 18 signatories (2010) out of the 35 required for ratification	• Equitable and reasonable use • Take account of all relevant social, environmental and economic factors • Take all appropriate measures to prevent the causing of significant harm to other watercourse states – the 'no harm' rule • Individually and jointly protect ecosystems • Cooperate for mutual benefit and good faith • Enables establishment of management mechanisms, info exchange and time-bound steps for conflict resolution based on fact-finding and impartial assessment.

Millennium Declaration (2000)	Millennium Development Goals (MDG) are 8 international development goals including poverty, health and education, gender equality, environmental sustainability.	All 193 UN member states and 23 international organisations have agreed	• Sustainable development; reduce biodiversity loss; reduce total water resources used • Halve the proportion of people without access to safe drinking water and basic sanitation by 2015
The Johannesburg Declaration and Plan of Implementation (2002) held at World Summit on Sustainable Development (WSSD) in Johannesburg	Plan of Implementation set a target for preparation of IWRM and Water Efficiency Plans	Non-binding; Recommended to General Assembly to endorse plan of implementation	• Sustainable development • IWRM; a river basin as a management unit • Biodiversity • Restoration of fisheries • Good governance • Partnership between governments, business and civil society

cross international boundaries, the challenges to IWRM and water resource planning are much more complex. Hence inter-state cooperation is of crucial importance.

While cooperation rather than conflict over transboundary watercourses is the norm (Giordano and Wolf *et al* 2003: 165), almost 90 per cent of all conflict events relate to water quantity or infrastructure such as dams and irrigation (Wolf *et al.* 2003). Examples of countries involved in contested rights and management include: Israel, Lebanon and Syria regarding the Jordan River; Turkey, Iraq and Syria regarding the Tigris-Euphrates; Nigeria, Niger and Mali regarding the Niger basin; Argentina and Uruguay regarding the La Plata, to name just a few (Rieu-Clarke and Loures 2009). Institutional arrangements between countries have been found to make a difference. For example, basins with high dam densities that have treaties have more cooperative relationships than those without (Giordano and Wolf 2003).

According to a recent study, only one third of transboundary basins have treaties, basin commissions or other forms of cooperative arrangements which apply to shared resources situated along the borders of two or more countries (UNEP 2006) (see Figure 2.1). The good news is that more than 54 new bilateral and multilateral water agreements have been signed since the Rio conference. Many states that did not support the still unratified International Watercourse Convention, such as China, have formalised bilateral agreements with neighbouring states.

Many agreements establish joint water commissions with decision-making, advisory and/or enforcement powers. Table 2.2 outlines some well-known multi-party agreements and the implementing bodies.

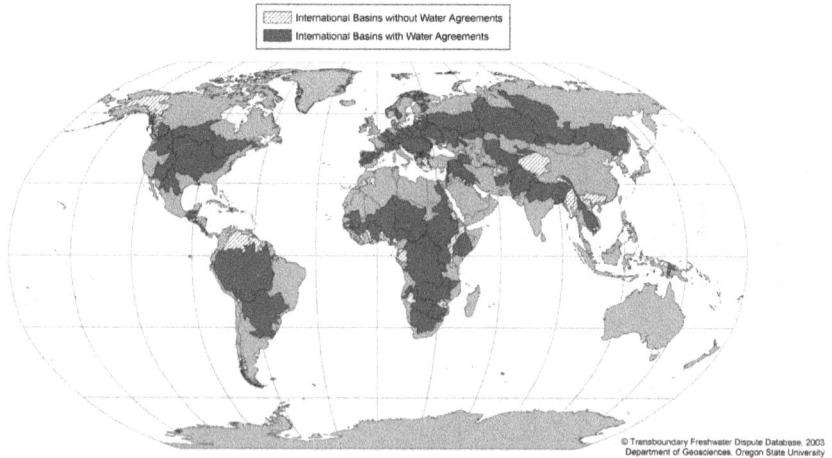

Figure 2.1   International river basins with agreements

(Source: Giordano and Wolf 2003. Product of the Transboundary Freshwater Dispute Database, Department of Geosciences, Oregon State University. Used with permission. Additional information can be found at: www.transboundarywaters.orst.edu.)

Table 2.2 Subregional and transboundary agreements

Agreement (date) and implementing body	Theme	Coverage	Principles addressed
Boundary Waters Treaty (US and Canada) 1909	Primarily oriented to maintaining navigation	Lakes and river and connecting waterways along which the international boundary between USA and Canada passes	Establishes order of precedence for use of waters: domestic and sanitary; navigation and canals (navigable waters be 'free and open'); irrigation and hydropower
			Controls on diversion of waters from the Niagara River
			Specifies equitable allocation for irrigation and power for St Mary and Milk Rivers
			Reporting and resolving disputes
1944 Treaty between the US and Mexico on the Rio Grande/Río Bravo *International Boundary and Water Commission*	Provision for joint development and use	USA and Mexico regarding Colorado River, Rio Grande (Rio Bravo) and Tijuana River	Establishes preferences for joint use as: domestic and municipal uses; agriculture; electric power, navigation, fishing and hunting
			Development and operation of dams and channels
			Plans for flood control, hydroelectricity, irrigation
			Attention to sanitation
			Specifies amount of water to be delivered in each country
			Equitable distribution

Agreement (date) and implementing body	Theme	Coverage	Principles addressed
Indus Water Treaty (IWT) 1960 *Permanent Indus Commission*	Incorporation of previously established norms of the water allocation mechanism in customary international water law. 'Equitable utilisation' reflected in both the equitable allocation of Indus tributaries to the two sides (three each) and the fact that either Party can equitably use rivers allocated to the other party for domestic, non-consumptive uses and irrigation. However, allocation is based on tributary locations; water is neither allocated on a quantitative basis between the parties nor managed by an operating rule. The IWT is specific about which nation uses which tributary, but it does not create mechanisms to address issues not specified in the treaty per se, such as groundwater use, changes in flow due to climate change, changing domestic demand due to population increases or rainfall variability. What to do? Co-riparians on other transboundary rivers have used mutual declarations as a basis for coordinated independent development (not joint) that can benefit each side. Although the quantity issue is addressed by the IWT, the notion of a reserve could still be applied to water quality, i.e. to sustain a minimum level of water quality in the Indus River Basin. They could jointly monitor WQ and treat discharge to standards. (Rahaman 2009)	India and Pakistan; mediated by World Bank which is also a signatory	Integrated basin management Cooperation and information exchange Notification if plans to construct engineering works that would cause interference to the other party Consultation and negotiation Peaceful settlement of disputes but little detail on definition and scope of 'no harm' (Rahaman 2009) Good faith Missing: sharing of benefits derived from water

Treaty relating to Cooperative Development of Water Resources of the Columbia River 'Columbia River Treaty' (USA and Canada) 1964 *Permanent Engineering Board* (50:50 members from both countries) reports annually to BC Hydro and US Army Corps of Engineers and Bonneville Power	Canada, USA	Cooperative development and operation of the water resources of the Columbia River Basin for the benefit of flood control and power. Clarified rights and obligations between British Columbia (BC) and Canadian government, allowed sale of Canadian entitlement to downstream US power benefits. US paid Canada for benefits of flood control and Canada was granted rights to divert water for hydropower (Giordano and Wolf 2003).	Benefit sharing: Unconditional storage for flood control defined plus additional on-call and one-half of US electric generation to Canada
La Plata River Treaty and bilateral and trilateral partial coalition agreements 1969 *Intergovernmental Coordinating Committee for the la Plata Basin Countries*	Argentina, Bolivia, Brazil, Paraguay and Uruguay	Promote and coordinate joint management, development and preservation of the basin. Subsequent multilateral and bilateral treaties outline the specifics of economic investment, hydroelectric development and transportation enhancement. Led to significant economic gains from the treaty arrangements for all five riparian states. But these economic gains have also generated significant externalities – particularly, severe environmental degradation (Gilman *et al.* 2008).	Requires open transportation and communication along the river and its tributaries. Prescribes cooperation in education, health, and management of 'non-water' resources (e.g., soil, forest, flora, and fauna)
Mekong River Agreement 1975	Laos, Vietnam, Cambodia, Thailand	Equality of right, not as equal shares of water but equal rights to use water on basis of each economic and social needs (Giordano and Wolf 2003).	Reasonable and equitable utilisation, not to cause significant harm, cooperation and information exchange, notification, consultation and negotiation, peaceful settlement of disputes (Rahaman 2009)

Agreement (date) and implementing body	Theme	Coverage	Principles addressed
Uruguay River Treaty 1982 *Administration Commission*	Protect and preserve aquatic environment of Uruguay River and Rio de la Plata	Argentina, Uruguay	Prevent pollution Notification of possible effects on navigation, water quality, and use of water Management of land and groundwater for no significant impairment Information exchange on fishing activity Negotiations and dispute resolution by ICJ
Treaty of Peace between The State of Israel and The Hashemite Kingdom of Jordan 1994	Water quality and protection	Jordan and Israel – shared waters of the Jordan and Yarmouk rivers, and Arava/Araba groundwater	To undertake protecting, within their own jurisdiction, against any pollution, contamination, harm or unauthorised withdrawals of each other's allocations To monitor jointly the quality of water along their boundary
Ganges Water Treaty 1996	30 year water-sharing arrangement recognising Bangladesh's rights as a lower-level riparian to allow cooperation in harnessing the water resources	India and Bangladesh	Equitable and reasonable utilisation Not to cause significant harm Cooperation Information exchange Notification, consultation and the peaceful settlement of disputes

The Convention on Co-operation for the Protection and Sustainable Use of the Danube River (Danube River Protection Convention – DRPC) 1998 *International Commission for the Protection of the Danube River*	Sustainable and equitable water management, including the conservation, improvement and the rational use of surface waters and groundwater in the catchment area as far as possible Sustainable water management based on criteria of stable, environmentally sound development, directed to: – maintain the overall quality of life; – maintain continuing access to natural resources; – avoid lasting environmental damage and protect ecosystems; – exercise a preventive approach. Reducing the pollution loads of the Black Sea from sources in the catchment area	15 members: Austria, Bosnia and Herzegovina, Bulgaria, Croatia, the Czech Republic, Germany, Hungary, Moldova, Montenegro, Romania, Slovakia, Slovenia, Serbia, Ukraine, and the European Union	'Polluter pays' principle Precautionary principle Prevention and reduction of transboundary impact, e.g. pollution Cooperate in monitoring, assessment and reporting, and information exchange Mutual assistance Hazard control from accidents, e.g. floods, ice
Nile Basin Initiative 1999 *NBI Secretariat support for Technical Advisory Committee and Nile Council of Ministers*	To develop the River Nile in a cooperative manner, share substantial socio-economic benefits and promote regional peace and security. Nile River Basin Cooperative Framework Agreement 2005 announced rights of upper riparian states to use Nile waters – unlikely to be signed by Egypt and Sudan.	Egypt, Sudan, Ethiopia, Uganda, Kenya, Tanzania, Burundi, Rwanda, the Democratic Republic of Congo (DRC), Eritrea (observer), supported by World Bank	Equitable utilisation of, and benefit from, the common Nile Basin water resources Objectives are: a. To develop the water resources of the basin in a sustainable and equitable way to ensure prosperity, security and peace for all its peoples. b. To ensure efficient water management and the optimal use of the resources. c. To ensure cooperation and joint action between the riparian countries seeking win–win gains. d. To target poverty eradication and promote economic integration. e. To ensure that the program results in a move from planning to action.

Agreement (date) and implementing body	Theme	Coverage	Principles addressed
2000 EU Water Framework Directive	A framework for community action in the field of water policy commits member states to achieve good qualitative and quantitative status of all water bodies, ground and surface waters (rivers, lakes, transitional waters, and coastal waters) by 2015	Sweden, Norway, Finland, Denmark, Estonia, Latvia, Lithuania, Poland, Germany, the Netherlands, Belgium, France, Spain, Portugal, Italy, Czech Republic, Austria, Slovakia, Slovenia, Hungary, Croatia, Romania, Bulgaria, Greece	Introduces River Basin Districts managed according to River Basin Management Plans, setting objectives and updated every 6 years to: Prevent deterioration, enhance and restore bodies of surface water, achieve good chemical and ecological status of such water by 2015 at the latest and to reduce pollution from discharges and emissions of hazardous substances; Protect, enhance and restore the status of all bodies of groundwater, prevent the pollution and deterioration of groundwater, and ensure a balance between groundwater abstraction and replenishment; Preserve protected areas; Encourage participation by all stakeholders in the implementation; Include adequate pricing policies and incentives, and penalties.

| Protocol for the Sustainable Development of Lake Victoria 2003 *Lake Victoria Basin Commission (LVBC)* | Sustainable development and management of the Lake Victoria Basin in areas of water resources, fisheries, agricultural and land-use practices, including irrigation, wetlands, environment and wildlife | Burundi, Kenya, Rwanda, Tanzania, Uganda | Equitable and reasonable utilisation
Protection and conservation of the basin and its ecosystems, including conservation of wild or endangered fauna and flora, wetlands, and fisheries resources
Each state to have legislation re: EIA, environmental audits, dumping waste
Adopt measures to prevent ecological harm to neighbouring States
'Polluter pays principle'
Cooperation, reporting and dispute settlement |
| Convention on the Status of the Volta River and the Establishment of the Volta Basin Authority 2007 | Cooperation for the rational and sustainable management of water resources of the Volta River Basin and a better sub-regional economic integration | Benin, Burkina Faso, Cote d'Ivoire, Ghana, Mali and Togo | Promote permanent consultation tools among the basin's stakeholders;
Promote the implementation of IWRM;
Equitable distribution of benefits;
Evaluate planned infrastructure developments that impact the water resources of the basin;
Develop and implement joint projects and works; and
Contribute to poverty reduction, sustainable development and socio-economic integration of the sub-region. |

International law and conventions provide an overarching commitment to principles, but more detailed guidelines are relied on for clear implementation. For example, the Ramsar Convention employs 21 handbooks that provide guidance on matters such as river basin management, managing groundwater, participatory skills, and impact assessment. The European Union Water Framework Directive 2000 (EU WFD) represents a significant legislative achievement, committing 27 countries to deadlines to manage and protect aquatic resources, with more than 28 guidance documents to aid implementation, several directly relating to water allocation (Box 2.1).

Box 2.1: EU Water Framework Directive implementation

The EUWFD incorporates into a legally binding instrument the key principles of integrated river basin management: participatory planning and management based on hydrologic entities, river basin districts, rather than political boundaries; consideration of the whole hydrological cycle and pressures affecting it; and integration of economic and ecological perspectives into water management through consideration of environmental protection and sustainable economic development. The WFD is accompanied by a timetable for action with deadlines for achieving good status for aquatic resources. Management is by integrated River Basin Management Plans (RBMP) with a Programme of Measures, developed through public involvement and updated every 6 years. The RBMP is the framework for integrating different policies and sectors.

Initial milestones of the Framework and Floods Directive included:

2003	Transposition into national legislation; identification of River Basin Districts and Authorities
2004	Identification of river basin pressures, impacts and economic analysis
2006	Monitoring network
2008	Draft RBMP
2009	Final RBMP
2010	Introduction of pricing policies
2011	Preliminary flood risk assessment
2013	Flood hazard and risk maps
2015	Flood risk management plans
2015	Achieving good status of surface and groundwater
2016–2021	The next cycle of RBMPs

A Groundwater Directive came into force in 2006, an Environmental Quality Standards Directive in 2008, and a Floods Directive in 2007.

Topics of key guidance documents include:

- Ecological assessment and Monitoring – No 6, 7, 10, 13, 14, 19, 25, 28
- Public participation and planning – No 8 and 11
- Groundwater – No 15, 16, 18, 26
- Changing climate and risk assessment – No 24, 26

As of 2012, 25 member states plus Norway had adopted and reported 121 RBMPs for their national parts of River Basin Districts (RBDs), out of a total of 170. Eighteen out of 23 had satisfactory flood risk management plans lodged by March 2012 (see Figure 2.2).

The EU WFD website has fact sheets, implementation reports, and studies such as attitudes towards water issues. http://ec.europa.eu/ environment/water/water-framework/index_en.html.

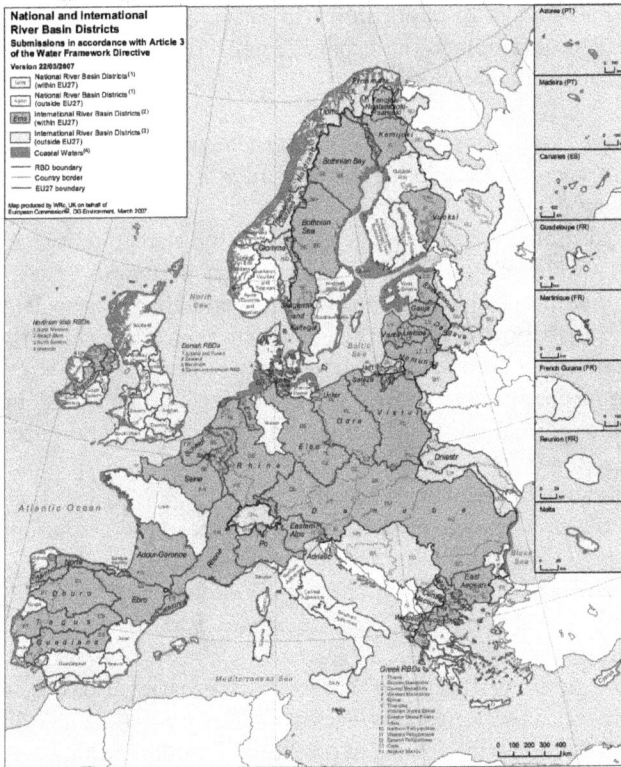

Figure 2.2   River basins in the European Union

(Source: European Union at http://ec.europa.eu/environment/water/water-framework/facts_ figures/pdf/2007_03_22_rbd_a3.pdf. © European Union.)

While the EU WFD is comprehensive in breadth of policy and depth in terms of guidelines, informed commentators suggest that many other international agreements are not sufficiently robust. A growing percentage of agreements deal with water quality, data exchange and information sharing, monitoring, and conflict resolution (Rahaman 2009), but a major area of conflict, water allocation, is seldom addressed in water accords. Those that do, frequently specify fixed amounts, not allowing the flexibility to deal with hydrological variation and changing values and needs (Giordano and Wolf 2003: 168). Equitable use, environmental protection, enforcement mechanisms, collaborative arrangements and public participation are often absent (Giordano and Wolf 2003: 169; Rieu-Clarke and Loures 2009). If groundwater is addressed in agreements, it is usually with reference to shared groundwater bodies connected to surface water systems, i.e. unconfined aquifers (Giordano and Wolf 2003). Indicators for measuring achievement of goals and effectiveness of conventions are also often poorly addressed.

With intensifying competition for water, it is suggested that cooperative international management frameworks can offer 'plenty of common ground and a window of opportunity to foster coordinated and sustainable water resources development and management' (Rahaman 2009: 171). Such frameworks should incorporate:

- flexibility, with an adaptable management structure allowing for public input, changing priorities and new information and monitoring technology;
- clear and flexible criteria for water allocation and water quality management, with allocation schedules and water quality standards that provide for extreme hydrological events, groundwater and changing societal values;
- equitable distribution of benefits, with an overriding value to not cause significant harm;
- concrete mechanisms to enforce the treaty through an oversight body;
- perceptions of accountability to the community and environmental watchdogs;
- detailed conflict resolution mechanisms;
- joint institutional mechanisms that can prepare implementation programs, facilitate collaboration, information exchange, and consultation, and make recommendations for policy change (Giordano and Wolf 2003: 170; McCaffrey 2003; Rahaman 2009; Bourblanc *et al.* 2012).

One of the reported criticisms is that many international basins still lack an effective joint management structure. An OECD study (2012: 19) found that key challenges are 'institutional and territorial fragmentation, badly managed multi-level governance, limited capacity at the local level, unclear allocation of roles and responsibilities and questionable resource allocation'.

Much has been learned and documented about the effectiveness of water governance structures particularly through the EU WFD, which has undertaken thorough reviews and evaluation of progress. Research suggests that some typical criteria for successful water governance structures, or 'the competent authority' (in the case of the EU WFD) include:

- clear and effective alignment of objectives
- transparency and accountability of the institutions and decisions taken
- meaningful sectoral and stakeholder involvement
- adequate human and financial resource allocation
- adaptability of structures and policies to changing circumstances.

In addition, unique to the situation of using the whole catchment as the basis for management in cross-boundary situations is the dilemma of pragmatic implementation under one organisation versus retaining separate responsibilities under different institutions. The latter is the usual case with transboundary watercourses where each country needs to retain its decision-making power and has its own division of jurisdictional responsibilities. This makes it even more important that sound coordination occurs:

- through multi-level governance and vertical integration of institutions to ensure that existing water management agencies and other actors within countries are not disenfranchised by an overarching basin-wide body and that the political environment can be adequately managed (Keskinen and Varis 2012; Karlsson-Vinkhuyzen 2012; Bourblanc *et al.* 2012);
- among agencies within countries so that inconsistencies due to delegation of responsibilities are addressed (EC staff 2012);
- to assist with developing capacity at the lowest appropriate level as per Agenda 21, as it is ultimately at the ground level that sustainable development will succeed or fail (Le Blanc 2012: 2).

Critical to the task of a watershed management agency is being successful at coordinating and influencing the actions of different institutions (Green and Fernandez-Bilbao 2006). We argue that a sound water resource planning framework with a process jointly implemented by nations, particularly those with shared watersheds, would provide a common goal and agreed process for resolving issues and reducing the likelihood of disputes. An effective institutional approach would enable parties to learn together (i.e. 'co-learn') through: joint fact-finding about their shared ecosystem, examining how sectoral activities and policies impact the system, and joint problem solving in a spirit of pragmatic cooperation and trust (Duda and La Roche 1997).

While all of this gives us some indication of the kinds of processes and actions that support effective water governance, the approach needs to be tailored for the individual circumstances, different uses and users. A Global Forum on Water Policy and Governance brought together eminent experts in water governance in June 2009. They identified as a major research priority, the need for at least 10 to 12 independent, objective and reliable case studies of good governance that could form a community of good practices from which countries trying to improve water governance can learn. They could identify the enabling environment and critical factors that contributed to their success. Countries or cities could then choose which aspects of each good practice would be most suitable to incorporate in their own water governance strategies (Biswas and Tortajada 2010).

International conventions guide conduct between nations. Within nations core water management principles need to be adopted in legislation and policies to not only meet international obligations, but also to 'do the right thing' domestically. The next section reviews some prominent examples of country-level policies and legislation.

National agreements, legislation and policies

Since the early 1990s, there has been a rapid progression in development of policy by countries at the national level to address sustainable management of water resources. Some 'stand-out' national initiatives are set out below. Given our Australian experience and global recognition that Australia has taken comprehensive initiatives to address water resource planning (UNEP 2012), we briefly describe the vertical linkages and evolution in thinking about water management in Australia, with particular examination of the National Water Initiative and changing institutional arrangements for the Murray-Darling Basin. We follow up with another 'star' in the show: the South African *National Water Act 1998*.

Australia

The Rio Convention in 1992 reflected an enhanced standing of the natural environment in public policy. It triggered change in Australian environmental policy, delivered initially through an Intergovernmental Agreement on the Environment (1992) and Strategy for Environmentally Sustainable Development (1992), followed by strategies for water quality, biological diversity, water for ecosystems and agreed approaches to impact assessment over the years (Table 2.3). One of the key drivers was the inherent variability of the resource which has meant that 'human actions to produce a secure water supply have had a profound impact on natural ecosystems' (Crase 2008a: 257).

A cornerstone for advancing water resource planning was the Water Reform Framework in 1994. To foster a faster pace of reforms in 2004 the National Water Initiative (NWI) committed the State governments to a programme of actions, accompanied by incentive payments. Much clearer direction was given for water resource planning (see Box 2.2).

Box 2.2: Australia's National Water Initiative 2004 stimulates water resource planning (see Commonwealth of Australia 2004)

In signing up to the NWI, the state and territories agreed to undertake transparent statutory-based water resource planning (clause 23ii) using the best available information (clause 36). Plans are to:

1 define environmental and other public benefit outcomes and put in place management arrangements to achieve those outcomes (clause 37);
2 define resource security outcomes and water allocation and trading rules and adjust over-allocated and/or overused systems (clauses 37, 43); and
3 put in place mechanisms to manage risk and adapt to improved information and knowledge, including monitoring and reporting (clause 40).

The NWI defines 'public benefit outcomes' of water use and management as 'mitigating pollution, public health, for example, limiting noxious algal blooms, Indigenous and cultural values, recreation, fisheries, tourism, navigation and amenity values' (Schedule B(ii)).

In terms of the process for developing the water resource plans, the NWI commits jurisdictions to:

1 consult and involve the community, including Indigenous communities (clauses 52, 95); and
2 actively consider trade-offs between competing outcomes for water systems, using best available science, social and economic analysis and community input, and address impacts on affected entitlement holders and communities (clauses 36, 97).

Schedule E of the NWI provides a guide to the contents of water resource plans and planning processes (see Box 3.2).

At a national level, the National Water Commission (NWC) is responsible for assessing compliance with the NWI, and assisting the states and territories to implement the NWI reforms, as well as auditing the implementation of the Murray-Darling Basin Plan. Similar to the EU WFD, regular reviews of progress in implementation of the NWI – 'biennial assessments' – have taken place. Considerable effort has been made by the NWC on studies of key issues such as groundwater and mining to further policy development. Examples of these will be referred to in later chapters.

Australian state and territory legislation and policies

In Australia, as the colonies became states of a nation, they retained super-ordinate legal status over water resources (see Figure 2.3). In each state and territory of Australia (except Western Australia) water resource planning sits in a contextual framework of legislation, policy and other types of plans. In all cases there is a primary statute governing water resources that sets out, to varying degrees, the purposes and content of water resource plans, the processes for preparing and reviewing the plans, the effect or outcomes of the plans and the relationship of the plans to other statutory instruments. Generally the statutes have 'objects' or purposes that guide the adminis-tration of the Act as a whole, including the development of water resource plans. Some state Acts are quite similar and most incorporate principles of environmentally sustainable development and reflect national water reform priorities; see for example the objects of Queensland *Water Act 2000* and the New South Wales (NSW) *Water Management Act 2000* (Table 2.4). All those performing a function or exercising a power under these statutes must seek to achieve the objects/principles/purposes.

One key difference between these two states is that NSW legislation recom-mends integrating water management with other aspects of the environment, on a whole-of-catchment-basis, whereas Queensland defines the type of waters covered such as rivers, aquifers and lakes. NSW also acknowledges the shared responsibility between government and water users. In both pieces of legis-lation some of the 'purposes' are defined more clearly in later chapters of the Act.

Frequently legislation requires plans to consider or be consistent with specific related policies and plans. State jurisdictions in Australia have also taken a variety of approaches to providing a more detailed policy framework or guidance for water resource planning. Western Australia, for example,

Figure 2.3 Australian states

Table 2.3   Strategic national water policy and legislation – Australia

Strategic policy and legislation	Goals and themes
Intergovernmental Agreement on the Environment 1992	An agreement between the Australian government and all states and territories setting out responsibilities for a cooperative national approach to ecologically sustainable development and better environmental protection. Committed to precautionary principle, intergenerational equity, conservation of biological diversity and ecological integrity, improved valuation of environmental assets, 'polluter pays', 'user pays', and impact assessment (including cumulative effects, consultation, joint processes). Acknowledged impact of climate change and committed to a National Greenhouse Response Strategy. Establishes cooperative arrangements and protocols for addressing cross border and international issues, and defining roles of governments to reduce disputes. Key mechanism for embedding State and Territory commitment to the National Strategy for Ecologically Sustainable Development (ESD).
National Strategy for Ecologically Sustainable Development (1992)	Australia's response to the Brundtland Report and Agenda 21. Strategy defines ecologically sustainable development as: Using, conserving and enhancing the community's resources so that ecological processes, on which life depends, are maintained, and the total quality of life, now and in the future, can be increased. The goal of the strategy is: Development that improves the total quality of life, both now and in the future, in a way that maintains the ecological processes on which life depends. The core objectives are: • to enhance individual and community well-being and welfare by following a path of economic development that safeguards the welfare of future generations • to provide for equity within and between generations • to protect biological diversity and maintain essential ecological processes and life-support systems. The guiding principles are: • decision-making processes should effectively integrate both long and short-term economic, environmental, social and equity considerations • where there are threats of serious or irreversible environmental damage, lack of full scientific certainty should not be used as a reason for postponing measures to prevent environmental degradation • the global dimension of environmental impacts of actions and policies should be recognised and considered • the need to develop a strong, growing and diversified economy which can enhance the capacity for environmental protection should be recognised • the need to maintain and enhance international competitiveness in an environmentally sound manner should be recognised • cost-effective and flexible policy instruments should be adopted, such as improved valuation, pricing and incentive mechanisms

Strategic policy and legislation	*Goals and themes*
	• decisions and actions should provide for broad community involvement on issues which affect them. These guiding principles and core objectives are to be considered as a package. No objective or principle should predominate over the others. A balanced approach is required that takes into account all these objectives and principles to pursue the goal of ecologically sustainable development.
The Murray-Darling Basin Agreement (1992)	Provides for the integrated management of the Murray-Darling Basin. The purpose of the agreement (Clause 1) is: to promote and co-ordinate effective planning and management for the equitable efficient and sustainable use of the water, land and other environmental resources of the Murray-Darling Basin. Replaced the 1915 River Murray Waters Agreement. Included New South Wales, Victoria, South Australia, Queensland, ACT. In 2007 the Agreement was subsumed within the Commonwealth *Water Act 2007.*
National Water Quality Management Strategy 1994	The process involves community and government development and implementation of a management plan for each catchment, aquifer, estuary, coastal water or other waterbody. Includes use of high-status national guidelines with local implementation.
COAG Water Reform Framework 1994	Committed state and territory governments to implement a framework to deliver a more sustainable and efficient water industry through a range of initiatives including cost recovery water pricing, clearly specified water entitlements, legally recognised environmental water, enhanced water entitlement trading and institutional reform. Implementation was linked to state governments receiving 'competition policy payments' from the Commonwealth.
National Strategy for the Conservation of Australia's Biological Diversity 1996	Foundation of ecologically sustainable development; one of the three core objectives of the National Strategy for Ecologically Sustainable Development. Biological resources provide food, medicines and industrial products. Biological diversity underpins human well-being through the provision of ecological services essential for the maintenance of soil fertility and clean, fresh water and air. Provides recreational opportunities and a source of inspiration and cultural identity. Aims to bridge the gap between current activities and the effective identification, conservation and management of Australia's biological diversity. Primary focus is Australia's Indigenous biological diversity and includes goals and actions in the areas of conservation, integration of biodiversity conservation into natural resource management and management of threatening processes. With regard to water resource management, the strategy requires governments to ensure that the conservation of biological diversity is 'taken into account'.

Strategic policy and legislation	*Goals and themes*
National Principles for the Provision of Water for Ecosystems 1996	To provide policy direction on how the issue of providing water for ecosystems should be dealt with in the context of general water allocation and management decisions, to ensure that these decisions would be sustainable in the long term. Introduction of comprehensive systems of water allocations including the determination of clearly specified water entitlements, the provision of water for the environment and water trading arrangements, as part of 1994 COAG water reform framework. Provided direction on how the issue of water for the environment should be dealt with in water allocation decisions, specifically in relation to: • the definition of environment • the aim of providing water for the environment • methods of providing water for the environment • management of environmental water provisions.
The *Environmental Protection of Biodiversity Conservation (EPBC) Act*, 1999	Commonwealth Government environmental legislation with objectives to: • provide for the protection of the environment, especially matters of national environmental significance • conserve Australian biodiversity • provide a streamlined national environmental assessment and approvals process • enhance the protection and management of important natural and cultural places • control the international movement of plants and animals (wildlife), wildlife specimens and products made or derived from wildlife • promote ecologically sustainable development through the conservation and ecologically sustainable use of natural resources. Under the Act, approval is required from the Australian Government Minister for the Environment and Water Resources for any proposed action, including projects, developments, activities, or alteration of these things, likely to have a significant impact on a matter protected by the EPBC Act. The environment assessment process of the Act protects matters of national environmental significance including: • World Heritage properties • National Heritage places • wetlands of international importance • threatened species and ecological communities • migratory species • Commonwealth marine areas.
National Water Initiative (NWI) 2004	The purpose is for governments and the community to determine water management and allocation decisions [for surface water and groundwater] to meet production, environmental and social objectives (clause 36), and to provide for (clause 37) secure ecological outcomes by describing the environmental and other public benefit outcomes for water systems and defining the appropriate water management arrangements to achieve those outcomes, and resource security outcomes by determining the shares in the consumptive pool and the rules to allocate water during the life of the plan.

Strategic policy and legislation	*Goals and themes*
	It commits to: • identifying overallocated water systems, and restoring those systems to sustainable levels • the expansion of the trade in water resulting in more profitable use of water and more cost-effective and flexible recovery of water to achieve environmental outcomes • more confidence for those investing in the water industry due to more secure water access entitlements, better registry arrangements, monitoring, reporting and accounting of water use, and improved public access to information • more sophisticated, transparent and comprehensive water planning • better and more efficient management of water in urban environments, for example through the increased use of recycled water and stormwater.
Commonwealth Water Act 2007	Subsumed the Murray-Darling Basin Agreement and established an independent Murray-Darling Basin Authority with the functions and powers, including enforcement powers, needed to ensure that basin water resources are managed in an integrated and sustainable way. Requires the Authority to prepare a strategic plan for the integrated and sustainable management of water resources in the Murray-Darling Basin – the Basin Plan. Establishes a Commonwealth Environmental Water Holder and provides for national collection, collation, analysis and dissemination of information about Australia's water resources and the use and management of water in Australia.

Source: Hamstead *et al.* 2008a

Table 2.4   Comparison of the purposes of Queensland and New South Wales water legislation

Queensland Water Act 2000	*NSW* Water Management Act 2000 (s 3)
• advance sustainable management and efficient use of water and other resources by establishing a system for the planning, allocation and use of water (s 10)	• apply the principles of ecologically sustainable development
• protect the biological diversity and health of natural ecosystems; • maintaining or improving the quality of naturally occurring water and other resources that benefit the natural resources of the State (s10(c)(iii)); • protecting water, watercourses, lakes, springs, aquifers, natural ecosystems and other resources from degradation and, if practicable, reversing degradation that has occurred (s10(c)(iv))	• protect, enhance and restore water sources, their associated ecosystems, ecological processes and biological diversity and their water quality

Queensland Water Act 2000	*NSW* Water Management Act 2000 (s 3)
• improving planning confidence of water users now and in the future regarding the availability and security of water entitlements (s10(c)(ii)); • economic development...in accordance with the principles of ecologically sustainable development (s10(c)(iii)) • recognising the interests of Aboriginal people and Torres Strait Islanders and their connection with the landscape in water planning (s10(c)(v))	• recognise and foster the significant social and economic benefits to the State that result from the sustainable and efficient use of water, including benefits to: environment, communities and industry, recreation, culture and heritage, and 'to the Aboriginal people in relation to their spiritual, social, customary and economic use of land and water' (s3(c)(iv)
• increasing community understanding of the need to use and manage water in a sustainable and cost efficient way (s10(c)(vii)) • encouraging the community to take an active part in planning the allocation and management of water (s10(c)(viii))	• recognise the role of the community, as a partner with government, in resolving issues relating to the management of water sources
• providing for the fair, orderly and efficient allocation of water to meet community needs	• provide for the orderly, efficient and equitable sharing of water from water sources
	• integrate the management of water sources with the management of other aspects of the environment
• efficient use of water – promotes demand management measures, water conservation and appropriate water quality objectives for intended use of water, water recycling, and 'takes into consideration the volume and quality of water leaving a particular application or destination to ensure it is appropriate for the next application or destination, including, for example, release into the environment' (s 11)	• to encourage the sharing of responsibility for the sustainable and efficient use of water between the Government and water users • to encourage best practice in the management and use of water

has specific policies addressing environmental water provisions, transferable water entitlements, water sharing and management of unused entitlements. Likewise, the Northern Territory has developed a water allocation framework that serves as a starting point for consideration of environmental water provisions in the development of water resource plans. Victoria, Tasmania and New South Wales have the three most comprehensively documented policy frameworks within which water resource planning operates. While Victoria provides a broad policy framework for water, the other two are a combination of policy and guidelines for water resource planning as outlined briefly below.

The Victorian state policy (Victorian Government 2004) is a high-level statement of the policy and processes for integrated water management and planning. It describes the state objectives and policy principles for environmental water and resource security and the mechanisms to give effect to principles such as plans and licensing. It sets the timeframes for bringing the new arrangements into place. It also sets the important policy of allowing the government, as a last resort, to adjust entitlements at 15-year intervals to restore river health. Within the 15-year period, other investment strategies will be used to achieve river health goals. It also sets the policy framework for compliance and monitoring.

In Tasmania, the document *Generic Principles for Water Management Planning* (DPIW 2009), was finalised in 2005. The principles provide the 'default' positions that will apply in developing water management plans, 'unless specific circumstances necessitate an alternative approach to achieve the dual aims of providing certainty to water users and protection of water resources'. The principles relate to planning processes, such as community consultation and plan review, as well as to the content of plans, such as objectives, allocation and licensing arrangements, trading, and metering and monitoring. They discuss the statutory and other policy requirements that must be met or considered when developing the actions and, in some instances, provide model provisions. The document is a very practical and easy to digest planning tool.

In New South Wales, the *Water Policy Advisory Notes for Water Management Committees* (NSW Government 2002) together with the statutorily recognised river flow and water quality objectives for individual catchments, the *State Groundwater Framework Policy*, and the *State Water Management Outcomes Plan* formed the policy backbone for the initial round of water sharing plans. *Water Policy Advisory Notes* were prepared on a whole-of-government basis to assist water management committees in developing water sharing plans. They were also used as an assessment tool when draft plans were submitted.

The *Water Policy Advisory Notes* were issues-based and outlined the issue, the government's position in relation to the issue and the rationale for that position (statutory or otherwise), the role of the plan in relation to it, and type of input sought from the management committee. Issues covered include:

- managing diversion limits in regulated and unregulated rivers
- supplementary water access
- floodplain harvesting
- high security water
- water extraction volumes and daily flow shares in unregulated rivers
- groundwater quantity management and dependent ecosystems freshwater flows to estuaries and coastal waters integrating water quality and river flow objectives into water sharing plans
- conservation of aquatic and riparian biodiversity and threatened species management
- Indigenous issues and cultural heritage protection.

These policy documents were also accompanied by standard templates for statutory water sharing plans.

More recently in New South Wales, the document *Macro Water Sharing Plans – The Approach for Unregulated Rivers* provides a complete policy and process manual for 'macro' water planning, generally applied to rivers from which there is less extraction and less competition for water. This comprehensive public document lays out the entire policy rationale and approach (NSW OofW, 2011).

A comprehensive overview of each State's legislation and policies is given in Hamstead *et al.* (2008b), Tables 2 and 3. Further information, regularly updated, is also available in the most recent NWI Biennial Assessment (NWC 2011c). Progress of each State government in implementing water planning is available through an online search tool, *National Water Planning Report Card* (accessed November 2013 at http://archive.nwc.gov.au/library/topic/planning/report-card).

Australian inter-jurisdictional arrangements

Similar to international cross-boundary agreements, where water management crosses jurisdictional boundaries between states within Australia, special arrangements are usually put in place. These generally involve legislation in each affected jurisdiction reflecting an agreement entered into, and an ongoing joint jurisdictional management arrangement. Agreements are shown in Figure 2.4.

The Murray-Darling Basin (MDB)

One of the most well known examples of coordinated water resource management in Australia is for the Murray-Darling Basin. The Basin comprises

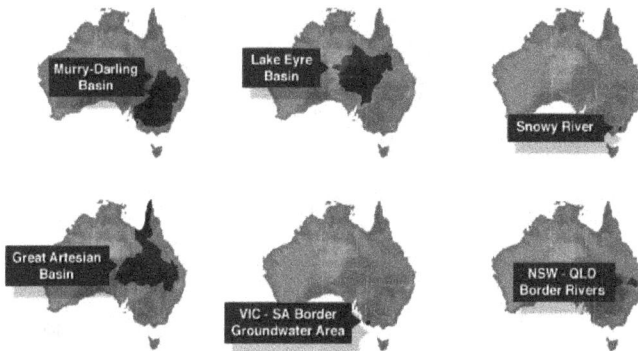

Figure 2.4   Australian inter-jurisdictional arrangements
(Courtesy of National Water Commission, Canberra, Australia)

over one million square kilometres and spans from the state of Queensland in the north, through New South Wales, the Australian Capital Territory, and Victoria, extending to South Australia in the southwest. Each State has its own history of, and approach to, development and regulation of water resources, making a coordinated approach to a shared watercourse somewhat problematic (Crase 2008b). After decades of discussion and negotiation, in 1915 an agreement was made about water sharing from the River Murray between the three southern states and the Commonwealth and given legal force by parallel legislation in each jurisdiction. A River Murray Commission was established to implement the agreement, consisting of a commissioner from each of the jurisdictions supported by a small bureaucracy. The River Murray Commission was essentially a water supply authority, administering the jointly funded sharing of water between the states and overseeing the construction and operation of dams and weirs to do so. Operation of the dams and weirs was contracted to state authorities. This cooperative arrangement proved effective for many decades in resolving inter-state water sharing and allowing cooperation in development of infrastructure to harness the water.

Commencing in the 1970s, it was evident that an agreement and institutions focusing on water supply were unable to address the growing challenges of declining water quality and condition of ecological assets. So in the 1980s a major institutional shift occurred that incorporated a more integrated catchment management focus. A new agreement was drawn up between the governments, which added to the existing water sharing arrangements for the River Murray mechanisms to investigate and address rising salinity in rivers and land across the whole Murray-Darling Basin, as well as a greater focus on river-dependent environmental assets. A new Murray-Darling Basin Commission was created with two Commissioners from each jurisdiction, with the second representative representing land and environmental management. This was supported by a larger bureaucracy and overseen by a new Basin Ministerial Council.

The first major success of the new institutional arrangements was the agreed implementation of a large-scale cap and trade scheme for saline inputs to the River Murray, that included allowing increases in saline inputs associated with development to be offset by salinity 'credits' resulting from works to reduce naturally occurring saline inputs. The second was the agreement to 'cap' or limit diversions of water from the rivers at a level consistent with the development existing in 1994, to preserve a proportion of the water flowing through the rivers for environmental and water quality purposes.

While the cap was highly successful in the short term, it was not completely so, as some aspects of water diversion escaped its implementation. The state of Queensland, though representing a small part of the Basin as a whole, held out for a higher level of development than that occurring in 1994, arguing it should not be penalised by being later to develop its resources. Additionally, the states struggled for years to develop mechanisms to limit some types of diversion, for example growth in storages collecting over-bank flow or overland runoff. Faced with continued evidence of environmental decline, governments entered into a new additional agreement (the Living Murray Agreement) in 2003 to jointly fund a program to 'recover' water lost to evaporation, or purchase water used for irrigation, and use it instead for iconic wetlands.

The millennial drought from 2003 to 2010 forced a further institutional rethink, as for the first time in many decades water shortages threatened the water supply of many towns and cities, irrigation water was unavailable, and iconic wetlands already lacking resilience due to degradation, faced complete destruction. Emergency measures were applied that over-rode long standing water sharing arrangements. In January 2007 the then Prime Minister of Australia announced an investment of $10 billion and an intention to seek a total change in institutions in the Murray-Darling Basin, with states transferring their powers to the Commonwealth to allow the Basin to be managed by a single water authority. After nearly a year of negotiation, a political change in Commonwealth government, and most states objecting to transfer powers to the Commonwealth, a new Commonwealth *Water Act 2007* came into force.

The Murray-Darling Basin Commission was dissolved and replaced with a new Murray-Darling Basin Authority (MDBA) reporting solely to the Commonwealth government. While states retained their constitutional powers to manage water, aspects of the Australian constitution related to trade and international obligations were used to allow the MDBA to prepare a Murray-Darling Basin Plan with the legal power to require actions within the states. The first Basin Plan became effective in early 2013, but preparation of the Basin Plan was highly controversial. Proposals for large long-term reductions in diversions across the Basin for environmental purposes were met with hostility by irrigators and rural communities. The whole interstate water sharing debate, relatively quiet for nearly a hundred years, was re-ignited, with the most downstream state, South Australia, pressing for more sizeable reductions, and upstream states arguing for less because of economic impacts. Assessments of environmental water needs, and social and economic impacts of reductions, were hotly disputed. The final Basin Plan addressed tensions to some extent by associating achievement of most reductions with investments to reduce 'losses', and deferring final implementation until 2019, during which time further review of the reduced diversion limits, and further negotiations about implementation are to occur.

Associated with the recent institutional changes, a role that has evolved is that of the environmental water manager. Under the Commonwealth *Water Act 2007* a Commonwealth Environmental Water Holder is established as a statutory role to manage the Commonwealth's environmental water entitlements to protect and restore the environmental assets of the Murray-Darling Basin, as well as outside the Basin where the Commonwealth owns water. States have similar entities that manage water owned by the state governments for environmental purposes.

What has been learned from the Murray-Darling experience? Institutional arrangements need to evolve over time to cater for changing circumstances such as increasing demand pressures and unexpected climatic events. Relying on cooperation and willing give and take to implement significant change among jurisdictions is overly optimistic; not only legislation but incentives are integral to achieving accepted outcomes. Jurisdictions and stakeholders are rarely readily willing to give up anything for a greater good that does not directly benefit them. A mixture of 'carrots and sticks' is usually needed to achieve change where any party stands to lose anything. Monitoring, based on solid data about biophysical and socio-economic characteristics and management actions, is recognised as crucial for adaptive management of the system. Tensions between productive use of water and environmental needs, and upstream and downstream users, are perennial challenges that can be resolved through water resource planning.

Republic of South Africa

Similar to Australia, South Africa has an uneven spread of water resources, highly variable climate and run-off, and flood and drought risks which affect communities as well as the national economy. It also needs to respond to the potential impacts of climate change. As the thirtieth most water scarce country in the world, South Africa has progressed in harnessing water resources to date in a challenging environment (DWA 2012). Yet, it still has issues with water quality and pollution, water security for social and economic development, and greater equity in delivering quality services across the population. An independent review found that water need not constrain growth and development in South Africa if effective water management is in place (Lenton and Muller 2009).

South Africa joined the global effort towards water reform in the 1990s. A substantial commitment is embedded in its post-Apartheid 1996 constitution which boldly 'guarantees everyone a right to have access to sufficient food and water and ... a right to an environment that is not harmful to health or well-being' (DWA 2012). A *National Water Policy* in 1997 was followed by the *Water Service Act 1997* and *National Water Act 1998* (NWA), as well as the *National Environmental Management Act 1998*. The NWA preamble and purpose highlight the intention to redress inequitable access to water.

Box 2.3: Purpose of South Africa's *National Water Act 1998* (s2)

To ensure that the nation's water resources are protected, used, developed, conserved, managed and controlled in ways which take into account amongst other factors –

(a) meeting the basic human needs of present and future generations;
(b) promoting equitable access to water;

(c) redressing the results of past racial and gender discrimination;
(d) promoting the efficient, sustainable and beneficial use of water in the public interest;
(e) facilitating social and economic development;
(f) providing for growing demand for water use;
(g) protecting aquatic and associated ecosystems and their biological diversity;
(h) reducing and preventing pollution and degradation of water resources;
(i) meeting international obligations;
(j) promoting dam safety;
(k) managing floods and droughts, and for achieving this purpose, to establish suitable institutions and to ensure that they have appropriate community, racial and gender representation.

The two pieces of water legislation promoted decentralisation of water management to the lowest possible level, recognised the role of stakeholder participation, and established Catchment Management Agencies (CMAs) governed by boards to represent major water users within respective Water Management Areas. Under the NWA, the Minister established guidelines for preparation of catchment management strategies (NWA s.10). The CMA must consult with the Minister, other relevant state agencies and any person or organisation whose activities might affect water resources and who have an interest in the CMA. The catchment management strategies must 'not be in conflict with the national water strategy' (s.9(b)). So getting this strategy right is vital.

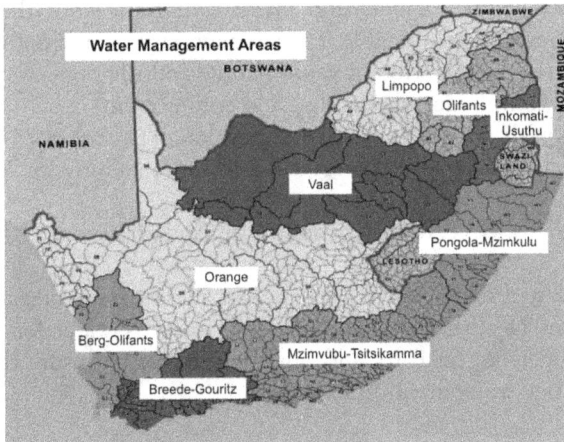

Figure 2.5 Water Management Areas in South Africa
(Source: DWA 2013: 65. Used with permission.)

South Africa's *National Water Resources Strategy* (NWRS) is the main means of implementing the *National Water Act* (NWA s5(1)). A review of the first Strategy, NRWS 1, found that reliable water supplies and infrastructure had improved and environmental flows had increased. However it raised issues such as limited implementation of water conservation and demand management and still inadequate redressing of past racial and gender imbalances in access to water for productive use. Better strategies for sharing water resources, the need for more equitable access to good quality water and need for involvement of water users have all been identified (van Koppen 2008). Commodification of water resources through full cost-recovery policies, corporatisation, and privatisation measures have been blamed for inequitable access to water (Gowlland-Gualtieri 2007).

According to the second Strategy – NRWS 2, South Africa has:

> sufficient water resources potential to meet its short to medium term requirements. But, the key challenge (and in order to avert a potential water crisis) is about mastering the art and science of unlocking the potential resources, ensuring timeous accessibility, facilitating sector and business viability (water cost and affordability), ensuring sustainable water delivery and management as well as effective water governance (DWA 2012: iv).

South Africa is not alone in identifying these critical success factors needing priority attention:

- financial resources and costs of water resource management
- governance, accountability, active water sector involvement, and partnerships
- a water management model and investment framework that includes a business approach, sustainable management, supply and delivery of water security with consideration of value change and life cycle
- regulation, compliance monitoring and enforcement
- alignment with national growth and development strategies
- skills shortage and need to invest in community and institutional capacity building
- improved knowledge, research, monitoring and evaluation (DWA 2012: v).

The NRWS 2 includes strategies to address these issues and has a pro-poor focus. In terms of allocation, the highest priority is water for the Reserve. The Reserve specifies quantity and quality of water to ensure that sufficient quantities of raw water are available for:

- the basic water needs of people who do not yet have access to potable water (at 25 litres per person per day)
- sustaining healthy aquatic ecosystems (DWA 2012: 40).

The next highest priorities, in order, are:

- meeting agreements with riparian states (international)
- water for poverty eradication, improvement of livelihoods of the poor and marginalised and uses that contribute to greater racial and gender equity
- uses that are strategically important for the national economy.

The NRWS 2 more recently reduced the original 19 CMAs to nine to improve viability and oversight. These will align with nine new Water Management Areas (see Figure 2.5).

What messages can we take from examining the Republic of South Africa Water Act? The critical success factors identified in South Africa have universal application. More broadly, similar to the Murray-Darling in Australia, governments need to regularly review and be willing to adapt institutional arrangements over time to achieve the goals; policies on prior-itisation of water access may be needed to ensure social and environmental objectives are not overwhelmed by short-term economic imperatives; and water resource planning needs to incorporate mechanisms aligned with objectives to ensure equitable distribution of costs and benefits.

Guidelines for water resource planning

Based on our analysis of challenges for water management (Chapter 1), conse-quent drivers for change, and the evolution of principles through treaties, interboundary agreements, and nation-level policies and legislation outlined in this chapter, we propose a set of guidelines for developing agreements, legislation and policies for water resource planning to maximise effectiveness in addressing the goals of IWRM – economic efficiency, social equity and environmental sustainability (see Box 2.4).

Box 2.4: Guidelines for establishing effective water resource planning

1 **Integrated and participatory governance for water resource planning and management**
   - Spatial scope based on hydrologic entities (e.g. river basins), rather than political boundaries; resource coverage including surface and groundwater (not just interconnected).
   - Thematic scope that includes the allocation of available water and the equitable sharing of the benefits that accrue from its existence and use and responsibilities for its management.
   - Commitment by governments and institutions (within and where relevant across jurisdictions) to multi-level and

---

multi-jurisdictional cooperation, supported by resources for planning and plan implementation commensurate with risks.

- Community engagement with and among water users, other interested stakeholders, and the general public including Indigenous communities to inform decisions, support conflict resolution, and support plan implementation.
- Accountability and transparency of institutions for a) information access and exchange with information provided in sufficient time for informed consultation; b) communicating reasons for decisions; and c) clear policy implementation and compliance rules.
- Adaptable management structure allowing for public input, changing priorities and new information and monitoring technology.

2 **Knowledge as a basis for decision-making**
- A good understanding of the water resource and its uses, future demands, risks and opportunities.
- Integrated consideration of economic, social and ecological perspectives when assessing risks and opportunities.

3 **Clearly defined and agreed desired outcomes and objectives to guide direction and strategies and actions for implementation**
- Development of agreed high level desired outcomes based around sustainable development and equitable sharing of benefits and costs, considering economic, social and ecological perspectives.
- Logical framework to identify actions to achieve objectives.

4 **Assessing options, risks, and impacts to achieve outcomes, mitigate harm**
- Active consideration of options for trade-off between competing outcomes, using best available science, social and economic analysis and community input.
- Inclusion of measures to mitigate impacts on ecosystems and communities.
- Inclusion of mechanisms to manage risk and contingency arrangements for extreme events.

5 **Monitoring and adaptive management**
- Ability to adapt to improved information and knowledge, including responding to results of monitoring and research.
- Adaptive management based on monitoring, reporting, and reflection on progressive implementation, environmental, social and economic characteristics of relevance, and effectiveness of management for corrective action as needed.
- A logical framework for water planning establishes indicators for measuring achievement of objectives and outcomes and effectiveness of management.

Despite the advances in water resources planning and management, it is still evident that the path of change is far from complete. In particular, population growth, extreme droughts and the threat of major ongoing climate change impacts mean the assumptions about sustainability that underpinned water resource planning in the past can no longer be used. We face a new reality, where risks are much greater, and the task of maintaining water ecosystems much harder and more costly than was considered possible a decade ago.

The next chapter outlines a framework for water resource planning that embodies many of the above guidelines. Chapters 1, 2, 4 deal with participatory governance issues; Chapter 5 identifies knowledge requirements; Chapter 6 covers setting objectives and a logical hierarchy for actions; Chapters 7 and 8 discuss identifying and assessing management options; and Chapter 9 expands on monitoring and the adaptive management process.

3 The planning framework

Scale and scope of water resource planning

In this book we focus on water resource planning as a process for the allocation and management of water from a defined water resource, which can be a river, a set of rivers in a catchment, the surface water in a catchment or part of a catchment, an aquifer or part of an aquifer, or combinations of these. The determination of the waters to be covered by the plan is normally guided by what is the most efficient and effective unit for management, subject to administrative limitations. Considerations include matters such as natural catchment (watershed) boundaries, hydraulic connections, legislative requirements and limitations, administrative boundaries (state/province or national), and location and intensity of water use development. Internationally there is a general trend towards planning at the river basin scale and inclusion of all surface and groundwater together to improve integration and holistic decision-making.

Box 3.1: Varying planning scales

The Padthaway Water Allocation Plan in South Australia is an example of a plan covering a relatively small area of approximately 700 km^2, with distances of 20 to 30km from one side to the other. Its boundaries are mostly straight-line administrative boundaries, with some parts defined by surface features. It incorporates a small town and surrounding rural farms with a population around 500. The plan deals only with groundwater use, with the average groundwater use in the plan area being around 40m m^3/year.

The Tana Catchment Management Strategy in Kenya is an example of a much larger river basin scale plan. It covers the Tana River Basin with an area of approximately 126,000 km^2 and a distance of approximately 500 km from the headwaters to the sea. It addresses both surface and groundwater, with a total estimated average extraction of 600 million m^3/year. The population in the area is around 7 million.

The 1,320 km long Rhine River in Europe runs from the Alps to the North Sea, with a catchment area of 200,000 km^2 spread over nine states. The complexity of the planning task has led to the plan being developed in two parts, an internationally coordinated overriding Part A plan (ICBR 2009) and Part B plans at national levels.

The Murray-Darling Basin Plan in Australia is a very large river basin scale plan. It covers the whole river basin with an area of 1.06 million km^2 and north to south span of around 1300 km. It addresses both surface and groundwater. The total population in the area is approximately 2 million and the average annual water extraction is around 12800 million m^3/year.

The scope of water resource planning also varies with respect to the management actions that it can command, for example whether it extends to demand management in cities and use of recycled water; land use activities that affect water, and so on. Where such matters are not included directly in the same planning and management system, they are often addressed through linkage arrangements such as overarching strategies, or requirements to consider effects outside the immediate scope of the plan.

Water resource planning is sometimes divided into two layers. The first is a broad plan (often called a 'strategy') that sets objectives over a longer term (10 to 50 years) and specifies broadly how those objectives are to be achieved. It might, for example, specify the volumes of water allocated for consumption and the environment, or the kinds of actions to be taken without being specific. The second layer comprises an action oriented plan (or plans) that aims to implement the broad plan through specifying sub-objectives and detailed actions. This may also differentiate allocations for different consumptive uses. For a single broad plan there may be multiple action plans covering parts of the area and/or parts of the timeframe of the broad plan. This is particularly useful in the case of complex jurisdictional arrangements, such as the Rhine River in Europe and Murray-Darling in Australia. The large Murray-Darling Basin Plan in Australia, for example, requires sub-plans (called 'water resource plans') at a river valley scale to be put in place to implement the diversion limits and other requirements set out in the Basin Plan.

To avoid over-complicating an already complex process, in this book we describe a single layered planning process, on the basis that multi-layered planning involves repeating the steps of the single process that we describe for each plan, with some apportionment of matters we describe within each planning cycle element between planning layers.

Water resource planning as a process

Water resource planning is essentially about weighing up the benefits of taking and using water for human water supply and economic purposes (e.g.

irrigation, industry, stock) on the one hand, against the benefits of leaving water in the water resource to maintain ecosystems and provide other benefits. It also considers the benefits associated with altering natural flow patterns (e.g. capturing high wet season flows and releasing them in drier times) in comparison to the need to retain sufficient natural patterns to support ecosystems. It entails planning how surface and underground water resources will be managed and shared to achieve environmental, economic and social outcomes. It is usually guided by the principles such as 'sustainable use' (refer to chapter 2), though what these mean in practice is frequently the subject of considerable debate. It is in the end what the local and broader community (extending to the global community) are willing to accept.

This agreed position is not determined solely by scientific assessments, though it is informed by them. It requires a process that involves judgements to be made on trade-offs between competing objectives and consideration of risk and uncertainty. How well this process is done determines how well the plan results in outcomes attuned to community needs and values in the short and long term.

Regardless of scale, water resource planning can be broken down into generic planning steps. These steps are described in a variety of different ways in different circumstances in the literature, but the underlying intent is common. For the purposes of this book, we label the planning steps with the general descriptive terms illustrated in Figure 3.1.

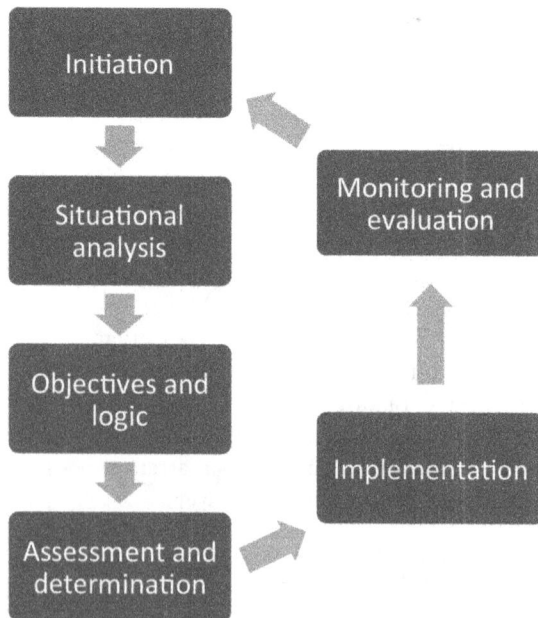

Figure 3.1   Steps of planning process

In simple terms, the core of the planning process is built around answering four questions:

- Where are we now? (situational analysis)
- Where do we want to be? (defining objectives)
- How do we get there? (determining logical linkages and actions)
- How will we know if we are on the right track? (monitoring and evaluation)

We add to this the steps of starting and framing the planning process (initiation) and implementation. We note also that adaptive learning (monitoring, and evaluation) can lead to revising the plan – essentially starting over, albeit from a more advanced position. In addition an over-arching part of the process is engaging stakeholders and the general public.

For simplicity, this process is shown as a series of sequential steps. In reality the process is considerably more iterative in nature, for example after assessing sets of possible actions there may be a need to go back and review objectives. Additionally the steps often overlap, for example with situational analysis and setting objectives occurring in parallel, and likewise with implementation and monitoring.

Revisions of plans are not always programmed. Frequently the aim is simply to create a plan in the first place, and little thought goes into how long it will remain unchanged and when and how it will be revised. This however makes some brave assumptions: that the plan will work effectively and efficiently; that the situation and available knowledge will not significantly change; and that the government and community's objectives will not change over time. For the purposes of this book, we assume that revision in one form or another is inevitable, and it is better to make adequate provision for it.

A short description of each of these planning steps is provided below. Subsequent chapters give in depth treatment.

Initiation

Initiation involves establishing the planning processes and organising the resources required to drive the process. Important aspects include:

- Defining roles and responsibilities
- Defining the scope and content of the plan
- Identifying issues and available knowledge
- Providing needed resources.

Initiating planning should also involve informing the public and stakeholders about the upcoming planning process and investigating the best ways to engage various sectors in the planning process (see Chapter 4). Obtaining initial feedback on issues and available knowledge informs the situational analysis.

Situational analysis

This step looks at the current status of the water resource, the benefits that the water resource provides (social, economic, environmental) and future threats, risks and opportunities.

A situational analysis provides a base for planning by clarifying current resource condition and use, and trends and factors that are likely to impact upon the resource in the future. It usually involves developing a model of the hydrologic behaviour of the system, which can be anything from a simple static model (e.g. a water balance showing estimated long-term average inflows, outflows and storage levels) through to a complex time-series model (e.g. a computer model that simulates the detailed behaviour of the system over many years on a short time-step). The situational analysis also attempts to estimate what future inflows and system behaviour might be, typically by extrapolating from the past. This is of necessity statistical rather than specifically predictive, with such future forecasts being in terms of likely long-term averages and trends, with likely statistical variations (e.g. daily/seasonal) around them. A range of possible future scenarios can be developed to allow for more robust planning for future uncertainty and risk associated with climatic variability and change.

Ideally, the situational analysis sets out how the behaviour of the water resource affects water-dependent ecosystems. This too can range from simple, broad conceptual models of cause and effect through to complex models of flows/water levels and ecosystem responses.

Situational analysis also involves developing an understanding of how much and for what purpose the water from the resource is currently used by people, e.g. for irrigation, domestic and town water supply, harvesting of flora and fauna for food and fibre, hydroelectricity, industry, recreation, cultural, etc. It also looks at likely future demands for water for these purposes. Amongst other things, this can lead to a more effective community engagement programme because it clarifies who the stakeholders are and the nature of their interests.

By providing an understanding of the cause and effect relationships and future opportunities and risks, the situational analysis can provide a foundation for the identification of actions and development of tools and indicators to assess the likely effect of actions on outcomes. Another use of the situational analysis is to provide a baseline for future monitoring of resource condition and use. In most cases, the monitoring program is established once actions are formulated and it is often part of the plan.

We discuss situational analysis in detail in chapter 5.

Objectives and logic

Given the situational analysis, this step is where broad decisions are made on the outcomes that are being sought. These represent the social, economic and environmental benefits and services that the government and community want from the resource into the future, and the water resource characteristics that support them. As shown later, we suggest an outcomes hierarchy consisting of non-quantified objectives and outputs, with associated quantifiable performance indicators and targets expressed in terms of those indicators.

Objectives are the foundation of a water resource planning process, reflecting the outcomes sought. They guide plan development and identify early agreement on common ground, which can be continually referred to throughout the process to gain perspective and keep moving forward.

Associated with the objectives, the broad logic of the plan can also be defined, setting out how the plan can contribute to achieving the objectives, and how success can be evaluated. This forms the basis for the next steps.

Often it is not possible to achieve all of the objectives because they are in competition with one another. For example, the taking of water for the objective of irrigated agriculture may result in failure to achieve other objectives relating to water quality and ecosystem condition. Thus the planning process has to resolve the level to which each of the objectives is to be achieved, or the 'weighing' of competing objectives through a 'trade-off' process.

Chapter 6 discusses how objectives and plan logic can be developed.

Assessment and determination

At this stage of the planning process, actions are developed on the basis of an understanding provided by the situational analysis and guided by the objectives and logic. The objectives and logic can constrain and prioritise possible actions.

Actions are typically packaged together into management options. For example one management option might be a certain sized dam, allocation rules and environmental flow requirements; another management option might be a different sized dam and allocation rules but the same environmental flow requirements. All sorts of variations are possible, but there are typically constraints such as whether new infrastructure is feasible or not. Possible actions are usually constrained by the institutional context, including such things as statutory powers, institutional capacity, available funding and overarching government policies. Subject to this, the identification of actions and management options may be done through an iterative process by an expert group, within government or in consultation with community advisors.

Comparative assessment of management options can be assisted by particular tools such as integrated impact assessments, socio-economic assessments, multiple-criteria or cost-benefit analysis, or simply interactive dialogue among agencies or with the community. This is the stage where trade-offs are considered:

- between benefits reliant on leaving water in the resource, and benefits reliant on extraction of water; and
- between benefits associated with extraction of water, either between different categories of uses (for example, water for urban use, mining and agriculture) or between users at different locations.

Determining the management option to be adopted should be based on consideration of the assessments of benefits and impacts and input from the community. It can be highly iterative, with options being adjusted or new options being developed as implications are investigated. It is often done in two parts. The first is a draft plan with recommended actions prepared by agency officers, that is the end result of their consideration of various management options. The draft plan is usually put out for public comment. The second part is a final plan that is put to a government Minister, or delegate with the appropriate authority, for approval.

Chapter 7 illustrates the kinds of actions that have been used for water management that could be considered in water resource plan development. Chapter 8 discusses how actions grouped as management options can be assessed and compared in order to determine what actions should be adopted in the plan.

Implementation

This is simply the implementation of the actions set out in the plan, typically by the agency responsible for managing the water resource in conjunction with stakeholders, either collaboratively or through regulatory or other mechanisms.

Monitoring and evaluation

No matter how much effort goes into planning or how good the science is, the reality is that no plan is likely to perfectly achieve its objectives because of challenges in implementation, uncertainty about cause and effect and unexpected changes in human water needs and climate. These factors mandate the need for a monitoring and evaluation regime. Regular 'formative' evaluations during implementation can improve plan implementation, guiding corrective action as required. Less frequent 'summative' evaluations that consider the effectiveness of the plan as a whole in achieving its objectives and the appropriateness of those objectives determine whether a revision of the plan is needed.

The design of monitoring programs and the evaluation schemes is discussed further in Chapter 9.

Delivering the process

To ensure agreement to a process and the resources necessary to carry it out, governments often embed the water resource planning process in policy and legislation. A couple of examples illustrate the range of matters covered and the degree of detail included. Following on from our discussion of Australia's national water policy in chapter 2, Schedule E to the *National Water Initiative* outlines a guide to the contents of water plans and planning processes, as shown in Box 3.2. In South Australia, legislation prescribes the processes that must take place in making a water resource plan, and the contents of the plan itself, as shown in Box 3.3.

Box 3.2: National Water Initiative (Australia) guide to content and process for water planning

1 Descriptions to include:
   - the water source or water sources covered by the plan (i.e. its geographic or physical extent);
   - the current health and condition of the system;
   - the risks that could affect the size of the water resource and the allocation of water for consumptive use under the plan, in particular the impact of natural events such as climate change and land use change, or limitations to the state of knowledge underpinning estimates of the resource;
   - the overall objectives of water allocation policies;
   - the knowledge base upon which decisions about allocations and requirements for the environment are being made, and an indication of how this base is to be improved during the course of the plan;
   - the uses and users of the water including consideration of Indigenous water use;
   - the environmental and other public benefit outcomes proposed during the life of the plan, and the water management arrangements required to meet those outcomes;
   - the estimated reliability of the water access entitlement and rules on how the consumptive pool is to be dispersed between the different categories of entitlements within the plan;
   - the rates, times and circumstances under which water may be taken from the water sources in the area, or the quantity of water that may be taken from the water sources in the area or delivered through the area; and
   - conditions to which entitlements and approvals having effect within the area covered by the plan are to be subject, including monitoring and reporting requirements, minimising impacts on third parties and the environment, and complying with site-use conditions.
2 Where systems are found to be overallocated or overused, the relevant plan should set out a pathway to correct the overallocation or overuse.

3 A plan duration should be consistent with the level of knowledge and development of the particular water source; and

4 In the case of ongoing plans, there should be a review process that allows for changes to be made in light of improved knowledge.

5 Further consideration to include:
   - relevant regional natural resource management plans and cross jurisdictional plans, where applicable;
   - an assessment of the level of connectivity between surface (including overland flow) and groundwater systems;
   - impacts on water users and the environment that the plan may have downstream (including estuaries) or out of its area of coverage, within or across jurisdictions;
   - water interception activities;

6 Water planning processes include:
   - consultation with stakeholders including those within or downstream of the plan area;
   - the application of the best available scientific knowledge and, consistent with the level of knowledge and resource use, socio-economic analyses;
   - adequate opportunity for consumptive use, environmental, cultural, and other public benefit issues to be identified and considered in an open and transparent way;
   - reference to broader regional natural resource management planning processes; and
   - consideration of, and synchronisation with, cross-jurisdictional water resource planning cycles.

(See Commonwealth of Australia 2004)

Box 3.3: Example of state legislation specifying the water planning process

In South Australia, the *Natural Resources Management Act 2004* sets out requirements for water allocation planning. The Act provides for the declaration of defined water resources (watercourses, surface water, lakes or water taken via wells) to be 'prescribed water resources' (s. 125). Once a water resource is prescribed, it triggers a series of actions leading to the regulation of water extraction by a licensing regime and the development and implementation of a water allocation plan (WAP) to set out how the prescribed water resources will be managed.

The Act sets out a process for developing a WAP which involves the following steps:

1 preparation of a draft concept statement
2 consultation on draft concept statement
3 finalisation of concept statement
4 conduct of investigations
5 preparation of draft WAP
6 consultation on draft WAP
7 adoption of WAP by the Minister.

The concept statement sets out the proposed content of the WAP, the matters that will be investigated prior to the drafting of the WAP, and how consultation will occur.

The Act requires the WAP to include:

- assessment of the needs of dependent ecosystems
- assessment of the effects on other water resources
- assessment of the capacity of the resource to meet demands
- water allocation criteria
- transfer criteria for water allocations
- matters to be considered in granting permits, for example, for wells or dams
- provision for monitoring of the resource.

A logical framework for water resource plans

Water resource plans have at their core a series of actions that are intended to achieve objectives. Inherent in these are assumed causal relationships between the resources to be applied (inputs), the implementation of the actions (outputs) and the achievement of the objectives. This can be called the 'programme logic', 'logic model' or 'logic map' of the plan.

Logic modelling or mapping has been used for many years in the field of program evaluation (McLaughlin and Jordan 1999). It is commonly used for evaluation of government programmes – to understand whether the programmes are achieving what was intended and are delivering value for money. A programme logic model shows how the programme will work showing the chain of relationships between resources, activities, outputs, outcomes and external influences. Associated with these is a set of performance indicators that can be used to evaluate the programme. Terminologies vary (e.g. goals, purposes, objectives, outcomes, etc.) but the underlying concepts are the same.

The Logical Framework Approach, originally developed for USAID during the late 1960s (see Coleman 1987, Team Technologies 2005) is a version of logic modelling that has been adopted by many international aid agencies for design and review of aid programs (see Table 3.1). The Approach maps a hierarchy of causal relationships into a 'narrative summary' with four elements: Goal, Objectives, Outputs, Components. For each of these elements there are corresponding performance indicators; means of monitoring and evaluating; and assumptions and risks concerning matters that are outside of the control of the program.

Whatever their form, logic models can be used:

- In the process of developing a water resource plan, to help to clearly conceptualise what is to be done and how it is to be achieved.

Table 3.1   Logical Framework Approach matrix

Narrative summary	Performance indicators	Monitoring and evaluation, supervision	Important assumptions
Goal Higher objective to which this project, along with others, will contribute.	Indicators to measure programme performance.	The programme evaluation system.	(Goal to Super Goal) Risk regarding strategic impact.
Development objectives The impact of this project. The change in beneficiary behaviour, systems or institutional performance because of the combined output and key assumptions.	Measures that describe the accomplishment of the development objective. The value, benefit and return on investment.	People, events, processes, sources of data for organising the project evaluation system.	(Objective to Goal) Risk regarding programme level impact.
Outputs The project intervention. What the project can be held accountable for producing.	Indicators that measure the value added of implementation of the components.	People, events, processes, sources of data – supervision and monitoring system for project implementation.	(Output to Objectives) Risk regarding design effectiveness.
Components The main component clusters that must be undertaken in order to accomplish the Outputs.	**Input / Resources** Budget by component. Monetary, physical, human resources required to produce the Outputs.	People, events, processes, sources of data and monitoring system for project design.	(Component to Output) Risk regarding implementation and efficiency.

Source: R. Moses Thompson, International President and founder of Team Technologies, Inc., in Team Technologies 2005: 16. Used with permission.

- To communicate to stakeholders what the provisions of the plan are for, and how they will contribute to the achievement of the identified outcomes.
- As a basis for evaluating the progressive implementation of the plan, to inform corrective action as needed.
- As a basis for triggering or informing a full-scale revision of the plan.

Unfortunately few water resource plans have explicitly stated logic models. This does not mean there is no logic in the plan. It rather means that the logic that was used to prepare the plan is scattered through the text of the plan and associated documents, or is not documented at all. It also means that there may be gaps or weaknesses in the plan logic that were never identified and performance indicators are more likely to be poorly designed. A common problem is plan indicators that are a 'grab bag' of disparate measures that are not systematically related to the plan logic, and prove to be of limited value for evaluation (GWP 2012).

Logic models can be expressed in a number of ways. Outcomes can be expressed in a hierarchy with broad outcomes at the end and one or more layers of intermediate outcomes. Figure 3.2 shows one way that logic models can be set out.

There can in reality be any number of levels of outcomes rather than just three, depending on how broadly the final outcome is defined, and how many intermediate levels are identified. For example, you could have the chain of outcomes broken into five levels as shown in Figure 3.3.

Considering this example, it is evident that the broader the outcome the more the achievement of the outcome is dependent on more than the single items listed. For example accessibility to groundwater also depends on there being appropriate infrastructure, and increased lifespan is also affected by public health measures. Some of the factors may be within the scope of the water resource plan, but many will not. The Logical Framework Approach explicitly documents significant externalities under the heading of 'assumptions and risks' (see Table 3.1) to make clearer the limitations of what can be expected from the programme or plan. It can be convenient to truncate

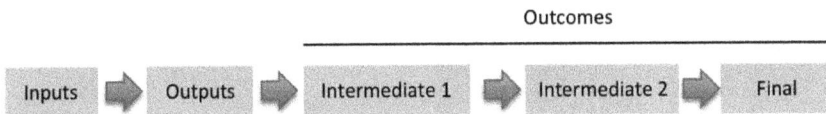

Figure 3.2   Generic logic model

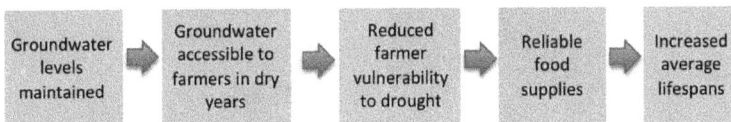

Figure 3.3   Example chain of outcomes

the documentation of the chain of outcomes in a water resource plan at some point, leaving the broader matters on the right to external state, national or international strategies to which the plan is subservient.

The chain of outcomes is not normally one dimensional, but rather is a hierarchy with several lower level outcomes contributing to higher levels. The Logic Framework Approach suggests drawing up this hierarchy (called a 'results hierarchy') then mapping it into the four levels of the narrative summary (see Figure 3.4).

A generic three level logic model developed for evaluating environmental water management in Australia is shown in Figure 3.5. The implied causal

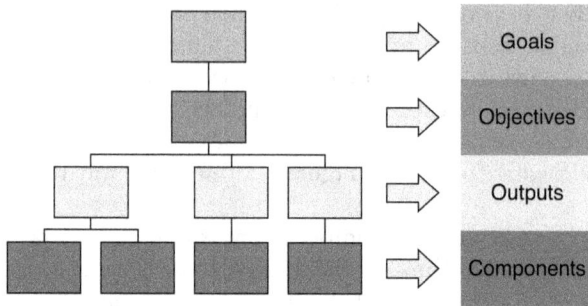

Figure 3.4   Mapping of results hierarchy into logic model

(Adapted from diagram by R. Moses Thompson, International President and founder of Team Technologies Inc., in Team Technologies 2005: 9.)

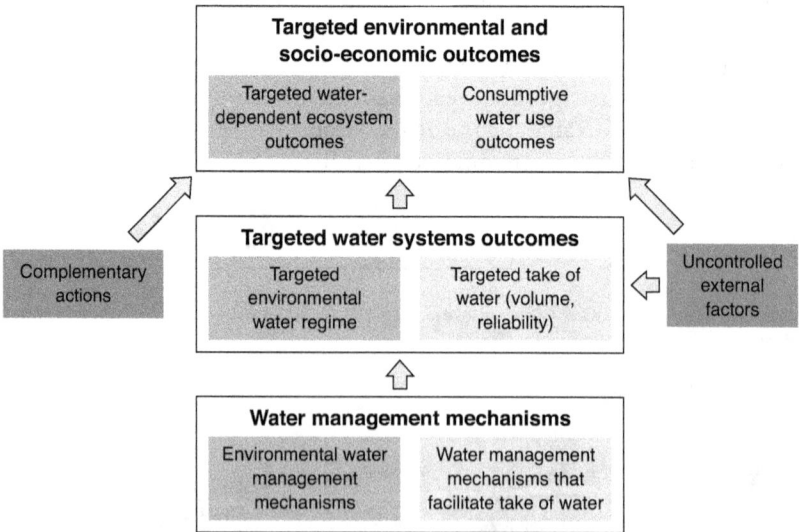

Figure 3.5   Generic logic model for environmental water management in Australia

(Source NWC 2011a. Used with permission)

Box 3.4: Logic model in the Katherine, NT, Australia, water allocation plan

The table of Outcomes, Objectives, Strategies and Performance Indicators in the Water Allocation Plan for the Tindall Limestone Aquifer, Katherine 2009–2019 (NT government 2009) provides an illustration of how one water resource plan has set out a logic model in a tabular form. Although the terminology and layout is different, it has similar elements to the World Bank (Team Technologies) Logical Framework. Part of this table is reproduced in Table 3.2.

Table 3.2 Extract from tabular logic model in the Water Allocation Plan for the Tindall Limestone Aquifer, Katherine, NT, Australia 2009–19

Outcomes	Objectives	Strategies	Performance Indicators
1. Ecosystems dependent on the Tindall aquifer, which are important for biodiversity, tourism, aesthetics, recreation and Indigenous cultural values, including springs and the Katherine and Daly Rivers, are preserved in good condition.	To preserve the following proportions of annual discharge from the Tindall aquifer to maintain base flow in the Katherine River: • During very dry years at least 87%; • During dry years at least 80%; • During normal or wet years at least 70% Protection of water quality within this water source and the Katherine River against degradation through extraction or bore construction	Annual extraction limits to be applied in accordance with Table 3, where the estimated unimpacted 1 Nov Katherine River flow is to be calculated at the beginning of each dry season using a model. Annual extraction limits in Table 3 may be adjusted following the review. To manage increases in extraction through water trading towards the Katherine River through water management zones. No new licences associated with bores able to take more than 20L/s within 100m of the Katherine River to be granted. Bores must not be drilled within 100m from potential sources of contamination. Bore construction permits will not be issued to properties that have access to reticulated water. Continue partnerships with research organisations to improve knowledge of ecosystem water requirements. Undertake consultation and research to improve understanding of Indigenous water issues and options to address them.	River health assessment parameters and ranges consistent with national guidelines will be developed in an implementation plan to this Plan. Annual discharge from this water source to the Katherine River relative to other years and annual extraction from this water source. Water quality in the Katherine River and Tindall aquifer. (Parameters and ranges consistent with national guidelines will be developed in an implementation plan to this Plan.) Identification of methodology to quantify water requirements for Indigenous cultural purposes. Identification of specific environmental water requirements that maintain ecological processes in the Katherine and Daly Rivers.

Source: Northern Territory Government 2009 pp.19–21. Used with permission.

relationships are: water management mechanisms (actions, regulatory rules, etc.) lead to water system outcomes (flows, water levels, water for extraction), which in turn contribute to broader outcomes (ecosystem health, economic production). The diagram also identifies that external factors also influence these two levels of outcomes, particularly the broader ones. Inputs are not separately specified in this model, but they could be added.

This three level logic model approach has also been recently applied to the evaluation of water resource plans in New South Wales (NSW), Australia. Thirty one water sharing plans (WSPs) commenced operation in 2004 and under state law expire in 2014, at which time they are either extended for a further ten years or remade. In preparation for this, the state water agency prepared evaluation reports of the plans, to assess whether change is needed and if so what should be changed. These plans as formulated did not include an explicit logic map, so the agency prepared logic maps for each plan by extracting information scattered through the plans and associated documents. These logic models were all based on a common three level model as shown in Figure 3.6.

Performance indicators can be designed for each level of the generic logic model shown in Figure 3.2. That is, there can be indicators for:

- INPUTS: were the inputs provided?
- OUTPUTS: were the outputs achieved?
- OUTCOMES: were the outcomes (at each level) achieved, or did the outputs contribute to the achievement of the outcomes?

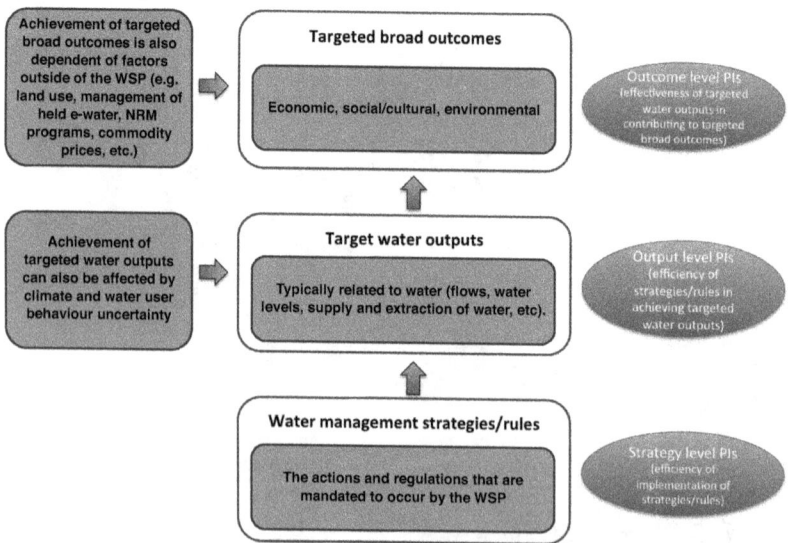

Figure 3.6 Common logic model for evaluation of NSW water sharing plans
(Source: Hamstead 2013)

The chosen indicators guide the appropriate monitoring program. For example if an indicator is groundwater level, then the monitoring program would include measurement of groundwater level at representative locations.

Selection of measurable indicators is critical to evaluation of the water resource plan. The indicators must be affordable and feasible to monitor, but at the same time clearly relevant to plan logic, otherwise they are a waste of effort and resources. Further discussion of performance indicators is provided in Chapter 9.

Logic framework terminology used in this book

As shown above, there are many variations in the shape of logic frameworks and the terminology used. For example, objectives are labelled in a myriad of different ways, usually reflecting terms used in local legislation, policy or practice, including visions, goals, purposes, aims, outcomes, outputs, principles, objectives, and targets. Regardless of the labels used, they are an expression of outcomes that are intended to be achieved either wholly or in part, in the short or the long term, and they can be expressed in a hierarchical logic framework, with different labels used for different levels. Likewise some plans label specific actions under the plan as strategies, rules, activities and other terms.

In the remainder of this book we adopt a terminology and logic framework for objectives based on an adaptation of the World Bank (Team Technologies) Logic Framework approach shown in Table 3.1, taking into account ecosystem services concepts from Plant *et al.* (2012) that are discussed further in chapter 5. This is shown in Table 3.3.

We define *objectives* as the desired benefits and associated services that derive from the water resource. *Outputs* are the water regime characteristics that are expected to deliver those benefits and associated services, subject to key external assumptions. *Actions* are the means that the plan puts in place to achieve those water regime characteristics. The objectives and outputs together are the *outcomes hierarchy* that is the core of the plan's logic.

For example, an *objective* might be improved economic production via irrigation and industry; corresponding *outputs* might indicate that water should be available to be diverted when and where needed for irrigation and industry; and the *actions* to achieve the outputs might include construction and operation of dams and weirs.

Objectives and *outputs* are mostly defined non-quantitatively. For competing objectives, the extent that they are achieved, or put at risk, will depend on the final actions adopted. *Performance indicators* and targets are used to indicate quantitatively the extent that objectives and outputs are expected to be achieved and thus the level of trade-off between competing objectives adopted in the plan after consideration of the benefits, risks and impacts of different management options. *Performance indicators* are the things to be measured, and *targets* are the levels of the performance indicators that the plan hopes to

Table 3.3   Logic framework and terminology used in this book

Level	Performance indicators and targets	Monitoring and evaluation	Key assumptions
Objectives The desired benefits and associated services resulting from a combination of outputs of the plan and key assumptions.	Indicators that define the extent that the objectives are intended to be achieved, and the effectiveness of the outputs in contributing to achieving them.	Means for measuring and evaluating achievement of the indicators for objectives.	Assumed actions and influences outside of the scope of the plan, upon which the achievement of the objectives is also reliant.
Outputs Desired water regime characteristics that produce the plan's contribution to the objectives.	Indicators that define the extent that outputs are intended to be achieved, and the effectiveness of the strategies in achieving them.	Means for measuring and evaluating achievement of the indicators for outputs.	Assumed actions and influences outside of the scope of the plan, upon which the achievement of the outputs is also reliant.
Actions The means that the plan puts into place to achieve the outputs.	Indicators of the extent/efficiency of implementation of actions, and whether/how they achieved outputs?	Means for measuring and evaluating achievement of the indicators for actions.	Assumed inputs and resources necessary to implement the actions.

achieve. For example, for the *objective* of improved economic production via irrigation and industry, a *performance indicator* might be production of rice in tonnes per year, and a *target* might be an annual rice production of 25 tonnes per year.

We acknowledge that in practice water resource plans may use different terms to these, and may not separate objectives, outputs and performance indicators in this way. However we argue that this is a useful framework that distinguishes the processes and underlying logic that actually occur. Practitioners need to consider what these elements are called in their plans and planning processes and make the necessary translations.

4   Consultation and collaboration

Fundamental to water resource planning is collaboration and engagement based on principles of good governance, identified in chapter 2. In this chapter, community engagement is defined; its purpose and role discussed. Methods for undertaking a stakeholder analysis, including use of community profiles and demographic statistics are supported by examples. This is used as the basis for developing a consultation plan which identifies purpose, methods, and a timeframe for consultation. The advantages and disadvantages of a range of consultation methods such as advisory committees are presented. Methods for engagement need to be tailored to the purpose of consultation and the needs of stakeholders, so effective ways of engaging Indigenous peoples, multiple cultures, or illiterate groups are discussed. Agency partnerships also need to be managed carefully, as valued stakeholders.

Community engagement as a principle of good governance

A commonly accepted principle for good governance of natural resources, including water, is 'inclusiveness' of key stakeholders and the broader community in decision-making. This principle is the case whether applied by international institutions such as the United Nations, World Bank or European Commission (Davidson *et al.* 2006: 8); found in government policy and legislation around the world; or in academic literature (Lockwood *et al.* 2010). While various terms are used for this, such as community, citizen or stakeholder participation, consultation or involvement, for the purpose of this book, we refer to it as 'community engagement'. We describe stakeholders as those who are affected by or have an interest in an action or decision. These definitions are explored later in this chapter.

In fact, community engagement provides a fundamental mechanism for achieving most of the other principles of good governance. For example, 'transparency' as a principle of good governance is achieved through ensuring a decision-making process is open for review by independent experts and that it withstands public scrutiny (Baldwin and Twyford 2007). Similarly, 'accountability' involves being able to give reasons for decisions and giving feedback about how community or expert input has influenced a decision. 'Fairness', another governance principle, is implemented through ensuring that those affected have a fair and impartial hearing. 'Adaptability' is more likely if key stakeholders are able to contribute practical advice and if open communication results in responsiveness to issues as they arise. All of these contribute to an agency's 'legitimacy' (yet another principle) as well as long-term relationships between a governance body and the community.

If good governance principles are not adhered to, there is great potential for water resource planning and management to generate conflict. It can take more time in the long term to achieve outcomes if one has to spend time reacting to negative public opinion. How many public servants have spent hours crafting letters that defend the government's position while ignoring the community's expressed concerns? This is the 'DAD' approach:

> Decide – the decision is made by the proponent or agency
> Announce – it is announced to the public
> Defend – it is defended in case of negative feedback

Such an approach rarely achieves better decisions yet increases cynicism and affects organisational credibility. A case in Bolivia provides an example (see Box 4.1).

Box 4.1: Lack of community recognition of water supply privatisation

In Bolivia, one month after the privatisation of water supply and sanitation services in the city of Cochabamba in 1999, Parliament adopted a law to provide the legal framework for sector regulation. The law did not include provisions concerning the recognition of the rights of Indigenous peoples and farmers. The privatisation of supply and the new law, combined with irregularities in the tender process, caused strong protests among the public against rate increases in urban areas without any prior improvement of services and the new legislation's effects on rural communities. Social unrest broke out in February and April 2000, followed by the declaration of a national state of emergency. The contract signed with the private consortium had to be terminated (Heiland 2009: 54).

The role of community engagement in water resource planning

Community engagement is a basic requirement of water resource plan preparation to ensure, at minimum, that both the needs of those directly affected and the values of the broader community are understood and considered. In addition, options can be tested to assess how they would meet stakeholder needs and values. By being involved in a water resource planning process, stakeholders gain an understanding of the complexity of water management, the vulnerability of water-related ecosystems, and of others' needs and values. From an elected representative's perspective, community engagement gives assurance that a wide range of views has been considered in the process. In most cases, stakeholder input has been found to improve the final outcome by mitigating undesirable effects, or assisting in finding a compromise between competing interests. Understanding the benefits of community engagement enables a water resource planner to negotiate for allocation of adequate resources (staff, funds for materials and travel) to the job required.

Community engagement can:

- improve the decision or outcome (more creative or flexible)
- inform people about the planning or decision-making process and the status of the resource
- build capacity and awareness among all stakeholders (including government and communities) about the resource and local and national agendas
- gain local knowledge of resources and use
- understand the range of values, concerns, and aspirations
- build relationships and partnerships
- seek alternatives and solutions
- identify and agree on appropriate criteria for testing options
- provide feedback on how public input influenced the decision
- engender greater acceptance of the decision with fewer implementation problems
- provide a litmus test for elected representatives when making final decisions
- resolve or reduce areas of conflict
- demonstrate an open government and enhance democracy
- contribute to social learning, triple-loop learning, and transformation (Mostert 2003).

From a water resource planning perspective, community engagement is particularly important for:

1 gathering information on issues, values, pressures, demands, possible impacts, and the nature of the resource; and
2 addressing procedural fairness in distributing benefits (and impacts).

Gathering information is important because not all the information needed for plan formulation is available in scientific assessments, and the community can be a valuable source of community-based local knowledge, anecdotal data, or monitoring data that might reveal history, trends or feasibility of options.

Procedural fairness is essential for successful plan making, particularly where there are tensions about how benefits or impacts are to be shared. It encompasses the following principles:

- All affected parties have the opportunity to hear and understand the potential implications of the plan for them. This means that stakeholders should be identified and the implications presented to them in a manner that is clear and understandable. This usually requires 'translation' of technical information into straightforward language.
- All affected parties have the opportunity to have their views presented and considered in decision-making. This requires both broad and targeted consultation strategies and ensuring affected parties know up front when and how they will have an opportunity to input. It also requires that the decision maker's response to submissions be recorded and reported.
- Decision-making should be, and be seen to be, unbiased and informed. This requires the technical information and assessments to use methods and be done by people who are seen to be unbiased experts, that the decision-making process is open and not able to be unduly influenced by particular groups, and that the decision-making person or body is perceived to be independent. The principles on which the decision is to be based should be predetermined and known.
- Decision-making should be open to cross-examination. This may be achieved by such things as independent review panels, hearings or capacity to appeal in the courts.

Transparency and openness is needed at all stages of the process.

Institutional foundation for consultation

As a result of the recognition that consultation is integral to good decision-making, it has been embedded in key international and national policies. For example UNDP policy is that 'all men and women should have a voice in decision-making, either directly or through legitimate intermediate institutions that represent their intention. Such broad participation is built on freedom of association and speech, as well as capacities to participate constructively' (Graham *et al.* 2003: 3).

Box 4.2: The Aarhus Convention

The Aarhus Convention (1998) – the Convention on Access to Information, Public Participation in Decision-making and Access to Justice in Environmental Matters – played an important role in building European consensus on community participation. By 2013, it was ratified by 45 states in Europe and Central Asia and the European Union. Its three pillars are:

1 Access to information: any citizen should have the right to easy access to environmental information. Public authorities must collect and disseminate information in a timely and transparent manner.
2 Public participation in decision-making: the public must be informed and have the chance to participate during the decision-making and legislative process about relevant projects. It acknowledges that people have knowledge and expertise to contribute and this can improve the quality of decisions, outcomes and guarantee procedural legitimacy.
3 Access to justice: the public has the right to judicial or administrative recourse where environmental law and the convention's principles are not adhered to.

Among other things, it specifies that regarding participation in plans and programs and legislation the responsible body should:

• involve the public early
• identify and enter discussions with the public
• employ reasonable time-frames for engagement
• allow public input in writing or a hearing
• should take account of outcomes of participation in decisions.

The major principles of the Aarhuus Convention have been applied in the EU Water Framework Directive (WFD), for example to 'ensure the participation of the general public including users of water in the establishment and updating of River Basin Management Plans (RBMPs – article 14). The WFD states that Draft RBMPs, which can be quite complicated, must be available for inspection and comment for one year before implementation.

A general commitment to basic stakeholder involvement – in the form of access to information or of providing for consultation with people affected is also anchored in international conventions or country constitutions (Ecuador, Brazil, Nepal). More-specific requirements for stakeholder participation are grounded in national legislation (Brazil, South Africa, USA, Nepal, New Zealand, Sri Lanka, Canada) and state or provincial legislation (Canada, Australia) (Baldwin and Twyford 2007). Australia's national policy, the NWI requires community engagement (see Box 4.3).

Box 4.3: Australia's National Water Initiative and requirements for community engagement

In Australia, the NWI requires 'open and timely consultation with all stakeholders' particularly in relation to pathways to returning systems to sustainable extraction levels; review of water plans; and decisions affecting security of water access entitlement or water use sustainability (s95). It requires that 'accurate and timely information' be provided to all relevant stakeholders regarding: implementation of water plans, trends in size of consumptive pool, and science underpinning environmental and other public benefit outcomes (s96). In addition, affected water users' communities and associated industry are to be consulted on possible responses to address adjustments to reduction in water availability (s97).

Schedule E (s6) of the NWI specifies that processes for water planning should include, among other things, 'consultation with stakeholders including those within or downstream of the plan area'. (Commonwealth of Australia 2004)

Having a legislative basis for community engagement in water resource planning is demonstration of a government's commitment to involving the community in decision-making. It ensures that at least certain minimum requirements are met, but needs to be flexible enough to address a range of situations. Effective consultation generally needs to go well beyond these minimum requirements. Detailed guidelines are frequently relied upon to provide specific direction for stakeholder participation (Nepal, Sri Lanka). Sometimes these guidelines are adopted as subordinate legislation (Water Resource WFD England and Wales Regulations); in other cases they have been adopted as policy principles. South Africa provides an example of a mix of somewhat limited legislated requirements but well-developed guidelines for consultation (see Box 4.4).

Box 4.4: South African consultation requirements

In the *National Water Act 1998*, South Africa requires the Minister to invite the public to submit written comments and to consider those comments on a proposed national water strategy (s5(5)(a) and (c)). Likewise the Act requires the same of catchment management agencies in preparation of a catchment management strategy or components (s8(5)(a) and (c)). The Act requires that the catchment management strategy must enable the public to participate in managing the water resources within its water management area (s9(g)) and take into account the needs and expectations of existing and potential water users (s9(h)).

Under the Act, the Minister may establish guidelines for the preparation of catchment management strategies. The consequent *Guidelines for the development of Catchment Management Strategies in South Africa* require that a catchment management agency must consult with any person or representative organisation 'whose activities affect or might affect water resources within its water management area; and who have an interest in the content, effect or implementation of the catchment management strategy' (DWAF 2007: 7). It calls for public involvement in creating a vision, advocates the 'right type of public participation at the right time' at all stages of the process, integrates engagement and capacity development, and suggests stakeholder participation in both CMS development and implementation. Under the Act water user associations can be established at a local level, as 'co-operative associations of individual water users who wish to undertake water-related activities for their mutual benefit' (Chapter 8 preamble), generally 'established around a single or multiple-use of water by licensed users' (DWAF 2007: 39).

International funding and development bodies, many of which have minimum requirements or guidelines, have also played an important role in bringing about effective stakeholder consultation. The World Bank requirement for participation in instances of resettlement – a result of the Nam Thuen 2 case – is an example. Efforts of NGOs such as International River Network have had a considerable impact on the adoption of improved practices. In addition, credit is due to the increasing number of corporate bodies that have adopted a corporate development ethic, signed legal agreements, or have issued guidelines referring to stakeholder participation or 'participatory development' (e.g., BC Hydro, Hydro Quebec, GTZ, and Meridian Energy (Baldwin and Twyford 2007)).

Definitions of community engagement

Many terms are used to describe public involvement in policy and decision-making processes, most commonly consultation, participation, and engagement. Community engagement tends to imply greater involvement of a wider range of people, and so is the preferred term in this book. Some definitions of community engagement are:

- Any process that involves the community in problem-solving or decision-making and uses community input to *make better decisions* (IAP2, 2003);

- A process through which stakeholders *influence* and share control over development initiatives and the decisions and resources which affect them (World Bank 1996, cited in Sidaway 2005: 119);
- A process by which public concerns, needs and values *are incorporated* into government decision-making (Creighton 1992, cited in Sidaway 2005: 119).

Two themes are common among these definitions:

- engagement involves the *community* (we discuss who later); and
- engagement *influences* making better decisions or improves the decision.

Community engagement should be used to assist when making difficult decisions, not just routine decisions. Making a better decision does not necessarily mean agreement, although that would be ideal. Twyford *et al.* (2006) states that community engagement should result in people feeling they have been heard, are comfortable with the process and can live with the decision. They suggest that for a decision to be better, it must be: informed; understood; implementable; and sustainable.

The International Association for Public Participation has identified core values for the practice of public participation as:

- being based on the belief that those who are affected by a decision have a right to be involved in the decision-making process;
- including the promise that the public's contribution will influence the decision;
- promoting sustainable decisions by recognising and communicating the needs and interests of all participants, including decision makers;
- seeking out and facilitating the involvement of those potentially affected by or interested in a decision;
- seeking input from participants in designing how they participate
- providing participants with the information they need to participate in a meaningful way;
- communicating to participants how their input affected the decision. (IAP2 2007).

Engaging the community in water resource planning processes can be time-consuming, requiring an extensive commitment in time and money by the government and the community. Yet it often saves time and effort in the long run. Key stakeholders in most democratic countries have an expectation that their views and input will be sought, respected and considered in the process. While it is usually made clear that a final decision rests with government (that is, government Ministers acting on advice from agency staff), stake-holders have an expectation that, in putting in the effort, they will in some way influence the outcomes. Government, too, expects that in committing to a process, stakeholders understand that they have a responsibility to consider all views and the broader public interest.

Having a sound approach to when and how to involve the community is a critical part of the water resource planning process, but it is important that it is designed appropriately. Good practice in initiating community engagement in water resource planning involves two steps:

1 preparing a stakeholder analysis, to identify who the stakeholders are and what their interests and issues might be in relation to the water resource plan; and
2 preparing a community engagement plan.

Who is a stakeholder?

A common question surrounds – Who are the stakeholders and the community? Who should be engaged? or Who has a right to have a say? Most engagement specialists take a broad and inclusive view of those who should be engaged.

The International Association for Public Participation (IAP2) defines 'community' as 'any individual or group of individuals, organisations or political entities with an interest in the outcome of a decision' (IAP2 2007).

We suggest that those who need to be included are those:

- whose work or life will be affected (e.g. might gain or lose economically)
- who live close to the location of a proposed initiative (proximity)
- existing and potential water users
- organisations and activities that might be affected
- whose customary habit, activity, route might be affected
- whose rights might be impacted (Indigenous peoples, minorities, those requiring equal access)
- whose values and interests may cause them to care about an activity
- with responsibility to make technical or policy decisions (i.e. government) (Baldwin and Twyford 2007; Sarkissian *et al.* 2003; Creighton 1992, cited in Sidaway 2005).

According to this description, then, anyone who has an interest, economic or not, should be able to have their views taken into account. Recognising that a wide range of stakeholders may be involved means that local and distant interests (or interests of the broader community) may conflict. Local communities have to live with the consequences of decisions being made in the broader public interest (Aslin and Brown 2004).

In particular, Sarkissian *et al.* (2003: 49) suggests using existing groups where possible but warns not to forget those with the smallest voice: those who are not, or not able to be, in an organisation due to ethnic or language background; income; literacy; working hours; age (young and seniors); gender or sexual orientation; single parents; or those with physical disability. In fact, it has often been said that the most important stakeholder is the one you leave out!

Stakeholder analysis

A stakeholder analysis provides the basic understanding of the characteristics and issues of the people likely to be affected or interested in water resource planning. Such an analysis provides the foundation for preparation of a Community Engagement Plan. Traditional techniques of consultation favour well-organised and well-educated people and groups with good communications skills. So it is important to also identify those who are at risk of being poorly represented and identify appropriate protocols and best ways to consult with them. It may require an introduction through a religious leader; men talking only to men, women to women; or initial engagement with custodians of particular land. In some cases, legislation or policy specifies categories of people to be involved. In Australia, the National Water Initiative pays special attention to engaging Indigenous peoples, in recognition of their past disenfranchisement in decision-making, and in the spirit of reconciliation (see Box 4.5). South Africa, too, acknowledges the need for redistribution and reconciliation.

Box 4.5: Australia's National Water Initiative requirements for Indigenous engagement in the water planning process

The National Water Initiative requires jurisdictions to provide for Indigenous access to water resources through planning processes, and to include Indigenous customary, social and spiritual objectives in water plans. Native Title interests are to be taken into account and Indigenous water use and interests assessed and addressed in plans.

Under the National Water Initiative there is an expectation that the environment and other public benefits will be identified as specifically as possible within water resource planning frameworks. According to the National Water Initiative, access by Aboriginal people to benefits related to water is to be achieved by water planning processes that:

- include Indigenous representation in water planning, wherever possible;
- incorporate Indigenous social, spiritual and customary objectives and strategies for achieving these objectives, wherever they can be developed;
- take account of the possible existence of native title rights to water in the catchment or aquifer area;

- potentially allocate water to native title holders; and
- account for any water allocated to native title holders for traditional purposes

(see Commonwealth of Australia 2004 clauses 52–54).

Understanding the social and cultural context is core to engaging Indigenous people in the water resource planning process. Examples are given later in this chapter and are discussed in Chapters 5 and 8 as well. Good sources about engaging Australian Indigenous people are Roughley and Williams 2007; Jackson 2008; Jackson and Altman 2009, given in the References.

Most thorough stakeholder analyses are based on preparing a 'social profile' by developing detailed descriptions of communities and groups of interest, using either secondary data (data already collected for other purposes such as ABS or local government statistics), or primary data from purpose-designed surveys, focus groups, meetings, interviewing community members, organisation staff, or group leaders. In the latter case first-hand information collecting may be the starting point of the community engagement process for the plan as a whole. The social profile could also become a baseline for use in further social and economic impact assessment (discussed in chapter 5). A typical social profile tailored to water resource planning might include:

- Details of local populations and specifically water users
  - Population trends, age profiles, education levels, employment, ethnic origin, average incomes, minorities, the disabled
  - Native Title and Indigenous water use and interests
  - Purpose for which water is used (horticulture, mining) and how it is used (e.g. type of irrigation)
- Local history
  - Early settlement, stories and key events related to water (e.g flooding; drought)
  - Trends in land and water use – distribution, development
  - Previous experience or history of consultation on natural resources and how that might impact current consultation
- Local industries and occupations in particular those related to water
  - Main employers (e.g. direct and indirect relationship to water such as food processors and farm equipment supplies)
  - Markets (e.g. local or distant requiring transport, long-term contracts or incidental)
  - Skills and services (e.g. local facilitation or chairing, sketching and design, printing)
- Relevant local issues and responses

- Issues, in particular water or resource based, in last 5 years; who/how got involved; what happened; how was each issue resolved?
- What are current issues and what is being done, e.g. planning studies?
- Attitudes to growth or resource use
- Organisations and key players
  - List principal groups, their activities and officers, their role in community
  - Identify community leaders and influencers, i.e. opinion leaders who can speak for each category of stakeholder
  - Knowledge and attitudes re water resource planning: local knowledge, myths, agendas, values, interests
  - Implications for communication, education
- Local communication channels (newspapers, radio, television, internet sites, newsletters, etc.)
  - Outline formal media, e.g. geographic coverage, capacity and credibility, circulation/audience
  - Identify informal networks – key nodes on grapevines
- Community services and facilities
  - Schools and colleges, health centres, libraries, meeting halls – can provide ideas about venues for meetings and appropriate timing of meetings so they do not clash with other events
  - Water infrastructure, rubbish disposal, sewerage services, irrigation schemes which support water use.

A stakeholder analysis can be initiated through a desktop analysis of media, government reports and statistics, and research articles about an area. It would then be followed up through interviews or meetings with those peak groups or influential people who are likely to be affected or interested in water resource planning. These people can suggest others who should not be left out – the 'snowball' effect. In particular, a stakeholder analysis should target information about water users (both extraction and others who benefit, e.g. tourists, recreational fishers) and water providers (e.g. urban water distributor) to understand use levels and patterns, investment, operational costs, level of knowledge about the resource, familiarity with water resource planning, concerns, values and interests. Understanding networks and linkages can assist in clarifying shared or common interests and power relationships.

Most experts on stakeholder analysis suggest identifying social networks and patterns of interaction; influence, power and authority; and interest in cooperating (Grimble 1998, Prell *et al.* 2009). Table 4.1 illustrates a simple way of summarising potential interests. Actual details will vary according to the particular situation. This can guide the method and frequency of consultation for particular groups, discussed further below.

Table 4.2 illustrates a more comprehensive matrix to indicate intention to collaborate, those in conflict with each other, or whether the relationship is uncertain; for example if water extraction needs to be reduced. This can assist in deciding how to group stakeholders for consultation and the methods used.

Table 4.1   Stakeholder analysis table

Stakeholder Group	Nature of interest or impact	Importance of interest	Influence
Water user, e.g. irrigator	Secure reliable source of water over a number of years	High	Medium to high
Conservationist	Water for environmental assets particularly at key times or seasons	High	Medium
Resident of town	Sufficient water for critical human needs (expected)	Low	Medium to high
Water manager	Smooth process with little conflict	High	Medium to high
Minster of Natural Resources	Best decision in public interest	Medium	High

Table 4.2   Stakeholder collaboration–conflict table

Stakeholder	1	2	3	4	5
1 (Government)		+	x	n	u
2 (Conservation)	+		x	n	+
3 (Irrigator)	x	x		+	u
4 (Business)	n	x	+		u
5 (City dweller)	n	+	u	u	

Key: + = collaborating; x = conflicting; n = neutral; u = unknown

A stakeholder analysis should identify how and where stakeholders would like to be consulted, at what points in the process they feel they can best contribute, and time frames to be considered (e.g. at a regular monthly meeting, not school holidays, not during rainy season or harvest). This provides input to the community engagement plan.

A stakeholder analysis can also be directed to provide a 'conflict assessment' when conflict about water resource planning is a possibility. This could be done as part of the stakeholder interviews or through confidential interviews by a neutral third party. It can be written up as a report protecting individual identity but clearly identifying stakeholders' issues and circulated to key stakeholders to facilitate basic exchange of views.

The community engagement plan

A community engagement plan, if developed and adopted early in the decision-making process, can be used to:

- gain commitment and agreement of decision-makers about the stages, purpose, time frame and human and financial resources allocated for participation;
- ensure transparency about the decision-making process for participants;
- select the level of participation and clarify participation goal, objectives and promise at each step, preferably in consultation with stakeholders;
- identify appropriate techniques for categories of stakeholders;
- identify and incorporate evaluation methodology at an early stage (Baldwin and Twyford 2007: 12).

The community engagement plan sets the stage for effective and efficient engagement with stakeholders and the community generally, and for doing so in a way that engenders community confidence in the planning process. A well thought out plan provides clarity about roles and realistic expectations for all participants. It also is a mechanism to gain agency agreement and commitment to the stages, purpose, time frame and resources (personnel, travel and materials) for a consultation process. The draft plan can be provided to possible participants for feedback on how they would like to be engaged, and whether stakeholders have been overlooked.

A common challenge in public consultation is matching community expectations of having influence on decision-making, with what can be delivered by the planning agency, given regulatory requirements and Ministerial responsibility for decision-making. This has resulted in considerable community disillusionment with water resource planning processes in the past. As a result, clarity is needed about what topics the agency is particularly interested in, if certain topics are off-limits or not open to discussion, and what areas of controversy can be expected.

Based on the stakeholder analysis, the community engagement plan considers the purpose of engagement at each stage of the planning process and determines which approaches to use to most efficiently achieve them. The IAP2 Spectrum is a now widely accepted model for selecting methods or tools according to the purpose and degree of engagement sought (Figure 4.1). Other models should not be overlooked. Arnstein's (1969) ladder of public participation is arguably the best-known framework used to illustrate degrees of public power and control of decision-making. It has been critiqued for assuming that those participation strategies higher on the ladder are better than those below them, without looking at the appropriateness of engagement, the degree of participation that is feasible and desirable given multi-layered governance, or willingness or capacity of stakeholders to take power and responsibility (Baker *et al.* 2007; Åström *et al.* 2011). It is a reminder, though, to seek to understand how communities can negotiate issues about power, to avoid co-option through ritualised consultation, and reconfigure local governance (Brownhill and Parker 2010). Other models to describe typologies or continuum of consultation can be found in Ross *et al.* (2002), Rowe and Frewer (2005), Sidaway (2005), and Sarkissian *et al.* (1999 and 2003). Basically they have a common message: the degree of involvement should be determined by the purpose of the engagement and level of interaction required. Additionally the level of investment needs to be proportional to the possible level of impact on stakeholders or the conflict

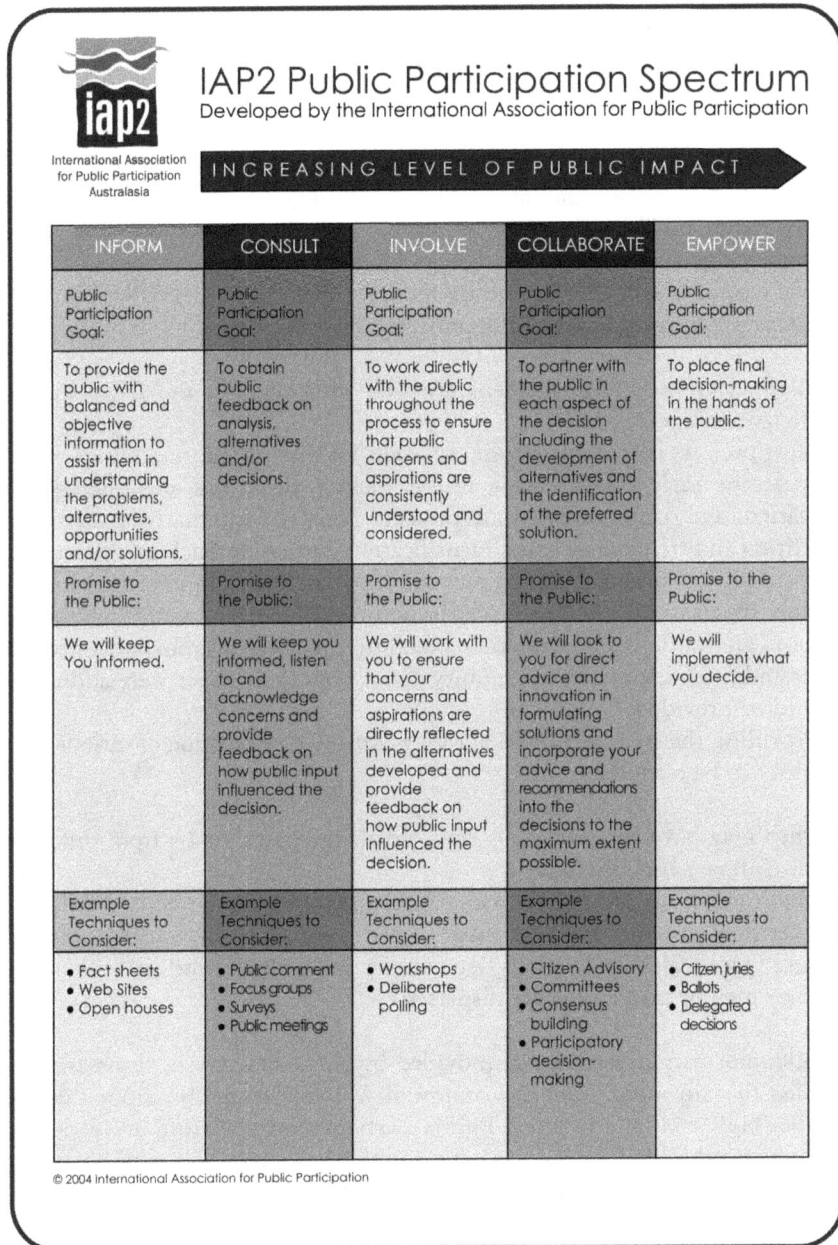

INFORM	CONSULT	INVOLVE	COLLABORATE	EMPOWER
Public Participation Goal:	Public Participation Goal:	Public Participation Goal:	Public Participation Goal:	Public Participation Goal:
To provide the public with balanced and objective information to assist them in understanding the problems, alternatives, opportunities and/or solutions.	To obtain public feedback on analysis, alternatives and/or decisions.	To work directly with the public throughout the process to ensure that public concerns and aspirations are consistently understood and considered.	To partner with the public in each aspect of the decision including the development of alternatives and the identification of the preferred solution.	To place final decision-making in the hands of the public.
Promise to the Public:	Promise to the Public:	Promise to the Public:	Promise to the Public:	Promise to the Public:
We will keep You informed.	We will keep you informed, listen to and acknowledge concerns and provide feedback on how public input influenced the decision.	We will work with you to ensure that your concerns and aspirations are directly reflected in the alternatives developed and provide feedback on how public input influenced the decision.	We will look to you for direct advice and innovation in formulating solutions and incorporate your advice and recommendations into the decisions to the maximum extent possible.	We will implement what you decide.
Example Techniques to Consider:	Example Techniques to Consider:	Example Techniques to Consider:	Example Techniques to Consider:	Example Techniques to Consider:
• Fact sheets • Web Sites • Open houses	• Public comment • Focus groups • Surveys • Public meetings	• Workshops • Deliberate polling	• Citizen Advisory Committees • Consensus building • Participatory decision-making	• Citizen juries • Ballots • Delegated decisions

© 2004 International Association for Public Participation

Figure 4.1 IAP2 public participation spectrum

(Source: International Association for Public Participation 2004. Used with permission.)

among stakeholders. For example, more resources might be needed for expert facilitation, information sharing, and dialogue where decisions involve major trade-offs.

Methods of engagement – techniques and tools

What are the best tools for engaging the community that would generate a good outcome? The choice of tools and techniques should be selected based on:

- purpose or objectives of having stakeholders participate: to share information; to gather and aggregate data; and/or to permit interaction (Baldwin and Twyford 2007)
- stakeholders' preferences and needs
- the resources available (including money, time, and skills)
- the size and complexity of the project.

Tools for purpose of engagement are well illustrated in the IAP2 Public Participation Spectrum.

Most projects include, at minimum, tools for sharing information, particularly at the early stages. These might include brochures announcing the initiation and time frame of a project; booklets summarising resource condition and trend or an issues identification document. Such information is particularly important to ensure a community can make an informed contribution. It can also be used to check the accuracy of data and understanding of issues. Later on in a water resource planning process, a summary of submissions and feedback to the community about how their issues were addressed should be provided.

Providing the right information to support the contribution of various user groups can be complex:

- they may have variable skills and knowledge bases, or be from different disciplinary backgrounds
- information needs to be accessible to the non-specialist
- they may have different information needs, requiring data at different time and space scales, and various degrees of prior synthesis and analysis
- they may be geographically dispersed.

While information needs to be provided by an agency, mechanisms are also needed to gain input from the community. Tools for public input that lie at the 'high end' of the IAP2 Public Participation Spectrum are expected to improve the likelihood of gaining public acceptance and making better decisions. They allow for greater interaction and deliberation. It can provide for input of local knowledge, especially important when there is little formally collected data. Though dialogue and deliberation cannot be expected to resolve all disagreement, they can often reduce areas of disagreement and offer opportunities for clarification, for identifying points deserving further analysis, and for encouraging mutual exploration of interests.

Use of advisory committees

In longer-term planning processes which may take two to three years such as for water resource planning, a common mechanism for effective and deliberative engagement is through a water advisory or consultative committee or working group comprised of key stakeholders. For example, in Baldwin and Twyford's (2007) review of good practices for engagement about dams, 10 out of the 13 case studies examined used consultative committees. In the Hamstead *et al.* (2008a) review of water resource planning processes around Australia, all processes had such committees.

The composition of these groups is important. Major water users such as irrigators or the mining sector are essential members as any change in allocation will have potential to directly affect them, and they are often a source of economic revenue and employment for the local region or the nation. However including other key stakeholders is essential to ensure balanced consideration of issues. They might include those concerned about domestic water supply such as a representative from a water delivery provider or from a town council and possibly a development aid group providing rural water. Those representing the health of a river and ecosystem and recreation or tourism interests should also be included. In many cases in Australia, Indigenous groups are invited to participate through a representative but this has not been found to be an effective way to gain insight into their interests (Jackson *et al* 2012).

The benefits of an advisory or consultative group is that they are: helping build mutual trust, respect and a collaborative approach; promoting open dialogue and the exchange of ideas; and generating shared understanding often through joint gathering and analysis of data (Baldwin and Twyford 2007). A community advisory committee can be an effective and efficient mechanism to allow for deliberation among diverse interests and to reassure and address concerns of interest groups before an issue gets misinterpreted or out of hand.

However it does not take the place of consultation with individual stakeholder groups and the broader community using deliberative techniques. Hamstead *et al.* (2008a) also documented cases from across Australia where committee processes have resulted in disillusioned community participants. The community, particularly advisory committees, may spend considerable personal time and effort participating in a deliberative process over time (e.g. a year or two). It therefore has a legitimate expectation that its views will not just be 'heard' but will contribute to decision-making. Disillusionment occurs when the final decisions are contrary to their recommendations, particularly where the reasons for this are not transparent, or are obviously the result of interest group lobbying outside of the process.

The risks of primary reliance on an advisory committee for engagement are several:

- not all interested parties belong to a group through which they can be engaged in an interchange of information;
- committee members may not be able, for a number of reasons, to provide adequate liaison with their constituents to build capacity and understanding or gain acceptance of a decision, without additional support;
- the circumstances of committee meetings may preclude meaningful and regular engagement by some groups (difficult access, work and family commitments);
- facilitation and mechanisms for equitable involvement within the committee and to outcomes are essential and require a skilled non-aligned facilitator;
- in committing to a long-term process, members want to see that they have some influence. They are disillusioned if the outcome is not one they agree with, in spite of knowing they are not the decision-makers.

So using a committee has its benefits and risks and requires a delicate balance of roles. Transparency, trust, and skill are key factors to making it work. Government interaction with such a committee can be used to identify topics and issues to be addressed more fully; and to develop approaches to address issues through broader consultation processes. Appointing members on the basis of expertise rather than representativeness may alleviate some of the expectations.

A good stakeholder analysis can help determine the best mechanism for engagement, and the need for, role and composition of a community advisory committee. Instead of a committee being the 'default' engagement approach, additional options for primary consultation may need to be considered, depending on the situation and who are the key stakeholders. Where a water advisory committee is formed it is important that there are clear terms of reference setting out members', and the government agency's or Department's, roles and obligations.

In many countries, such as Indonesia and Kenya, engagement by government is primarily with water user organisations who are not only those most directly affected by any changes in the water regime, but participate to some degree or other in co-managing the system. Described in more detail in Chapter 7, these groups are comprised of small-scale irrigators.

Other tools and techniques

Other successful mechanisms have been workshops held with specific interest groups: dedicated consultation officers available to interact, register concerns, and organise meetings; resource information centres with displays, and use of appropriate languages or translation services as appropriate.

Instead of relying on workshops, focus groups, websites, and advisory committees, some innovative methods can be used such as 'fishbowls', 'photo-voice', round tables, and open houses. The method of engagement needs to be tailored to the needs and background of participants and the purpose of engagement. The meaningful involvement of the Indigenous communities is a case in point, especially where the time frames and 'ways of doing business' are culturally different to non-Indigenous methods of committees, information gathering and delivery. For the most part, Indigenous cultural landscape (landforms, the natural environment) is integrated with beliefs and values not compartmentalised into water, vegetation and land management issues as practised by planners and government agencies. As a result their interests in water might be expressed quite differently than others.

Sources of information on a wide range of tools and techniques and their potential application are available on:

- Practitioner tools – International Association for Public Participation (IAP2 2003) – www.iap2.org
- Citizen science toolbox – https://www3.secure.griffith.edu.au/03/toolbox/
- Aslin and Brown's (2004) toolkit for engagement for Murray-Darling Basin – http://www.mdba.gov.au
- Tan (2013, 68) summarises engagement tools developed as part of the *Water Planning Processes: Lessons, Gaps and Adoption* project to address water challenges (Tan *et al.* 2010).

Methods are available for engaging large numbers, in an urban catchment or an interjurisdictional river system, for example (see Box 4.6). Online methods using information and communication technologies are becoming a standard tool, especially where there is a high penetration of internet users and good download speeds. Websites are a common method of providing information inexpensively, including colour maps. Online surveys can seek structured anonymous input. Increasingly, social media such as Twitter and Facebook are used for interaction, and agencies have developed guidelines for how to use them effectively and avoid risks (DPW 2010). For example social media might be used where there is an expectation of ongoing dialogue with government and to supplement other channels of communication. In general, information and communication technologies are intended to facilitate increased participation especially by those who may not be able to participate otherwise due to work or family commitments. Governments also intend that material provided online will provide increased transparency. One of the risks of relying solely on social media is alienating those who cannot easily access or make use of the internet or social media for whatever reason. Download speeds or intermittent electricity in rural and remote areas, for example, may discourage this type of access.

Special consideration is also needed for addressing different types of stakeholder diversity: disabled, culturally linguistically diverse, Indigenous people, older people, children, young people, and those who are disaster

Box 4.6: Tennessee Valley Authority (TVA), USA, Reservoir Operations Study

TVA is a multipurpose federal corporation operating the largest public power system comprised of dams and reservoirs with associated facilities in the USA, serving almost nine million people in parts of seven southeastern states. The system is managed to serve additional needs regarding flooding, river transportation, domestic and industrial water supply, recreation, and habitat. In 2001, the TVA initiated a comprehensive review of its reservoir operating policy, incorporating comprehensive public involvement. The TVA examined a broad range of policy alternatives, which would change reservoir levels and flow releases and their seasonal timing to produce a different mix of benefits. It included a long-range planning horizon to the year 2030. The process was concluded in May 2004 and endorsed in June 2004.

The community engagement programme began with "outreach" to identify the public's preferred reservoir operation priorities. TVA did an initial mailing to 66,000 Valley residents, established a web presence and toll-free phone line, and began targeted media outreach. At the same time, the team began work on the centrepiece of the public participation effort – community workshops organised to encourage public comment and participation from all interested parties and provide a longer-term opportunity for relationship building and two-way communications with stakeholders. The audience for the outreach effort included all members of the public who use the Tennessee River – power customers; commercial shippers; reservoir-user groups; property owners; local, state, and federal elected officials; tourism groups; environmentalists; and many other area citizens.

In 2002, the team managed 21 community workshops across the seven-state region to gain input on the desired scope of the study. The following year, to present specific policy changes considered in the study, an additional 12 workshops were conducted across the region. A special briefing preceded each of these 33 workshops for public officials and other local opinion leaders. An additional 47 briefings for public officials and other opinion leaders were conducted during those two years, and 24 such briefings were held in 2004 when the outcomes of the study were announced. More than 3,000 Valley residents attended the workshops, and thousands of others commented through the website and other means.

To encourage full participation by those attending the workshops, TVA contracted with a firm specialising in collaboration tools for an innovative, interactive system for encouraging and recording comments at the community workshops. The firm provided a computer system and facilitators for each workshop, along with follow-up reports documenting the public comments. At each workshop, participants were invited to join in an opening session to voice their preferences for river system operations where they recorded their opinions on how TVA should operate the river system using keypads (computerised multi-voting). Responses were tabulated and displayed electronically, enabling everyone to see the range of opinions within minutes. Laptop computers were used to comment during smaller, facilitated discussions. The laptops proved overwhelmingly popular with every demographic and their use far exceeded that of other vehicles for providing comments, including court reporters and cards for written comments. The result was an electronic database that reflected the views of everyone, not just a vocal few.

The briefings and community workshops featured full-colour displays, fact sheets, and videotapes, as well as subject-matter experts available for group and one-on-one discussions. In-person, e-mail, and mail updates to local leaders took place regularly, and four editions of a tabloid-sized newsletter went to interested citizens to introduce the review process, track its progress, and encourage their participation. Later in the study, as local issues and concerns emerged in one geographic region, TVA worked with local community groups to conduct additional sessions with the public and briefings with opinion leaders to address specific questions.

TVA also established two groups – a 17-member Interagency Team and a 13-member Public Review Group – to ensure that agencies and members of the public were actively and continuously involved throughout the study.

In June 2004 the Board adopted a preferred alternative that it felt established a balance of reservoir system operating objectives that was more responsive to values expressed by the public, while remaining consistent with the operating priorities established by the TVA Act.

The success of the public involvement process was evident in the decline in the number of complaints from stakeholders concerning TVA's operation of the river system. Some of TVA's most vocal adversaries have since become advocates, playing a lead role in informing others about operating constraints. Editorials in Valley newspapers have been positive, reinforcing several key messages: the credibility of the study, TVA's responsiveness to public opinion, and the need to manage the river system to balance multiple, competing needs. In addition, a comparison of opinion surveys conducted at the beginning and near the end of the study shows a significant improvement in public officials' perceptions of TVA.

This box includes extracts from a report by Baldwin and Twyford for UNEP (2007), reproduced with permission.

affected. These people might offer insight into history of water issues and options and strategies for minimising impact. For example, consultation about water for market gardens operated by Vietnamese-Australians in western Sydney, Australia may consider use of an interpreter or tapping into a particular social or religious network.

No matter what the information, it needs to be seen as credible by the community and defensible by scientists and independent reviewers.

Box 4.7: Tailoring visual methods to interpret science to stakeholders

Baldwin *et al.* (2012) report on a project that trialled two different visual methods to assist communities to understand groundwater characteristics in their areas. Visual images have been found to rapidly increase people's environmental awareness and stimulate public engagement. Using visualisation to portray groundwater enables users to 'see the unseen' (ibid. 76). A preliminary stakeholder analysis revealed the different needs of two communities.

A relatively sophisticated group of irrigators in the Central Condamine Alluvium region in Queensland needed to accept a large reduction in water allocation during the water planning process, which would result in a cost to business. To build their trust and credibility in the data, Groundwater Visualisation System software was used to develop a tool to show a simple display of the time/space variations in groundwater hydrology. Irrigators were given a CD and taught how to manipulate the model image by zooming in and examining cut cross-sections of the aquifer. The images used historical data to show the effect of extraction on the level of the water table (Cox *et al.* 2013). Participants thought it a useful tool for presenting information to less involved stakeholders as a stimulus to informed discussion. See Figure 4.2 in colour plates.

The second visual tool was used in advance of a water planning process in an undeveloped groundwater area in the Tiwi Islands in Northern Territory, Australia. Residents were unfamiliar with groundwater use. An interactive 3D physical groundwater model was constructed in a plexiglass box that circulated water through a cross-section of soil to demonstrate relationships between groundwater, rainfall, aquifer recharge, production bores, billabong, and spring flow. Being battery-operated it could be used in remote areas. It did not rely on written information and was a good catalyst for provoking discussion about on-ground features. See Figure 4.3 in colour plates.

Box 4.8: Community mapping, 'Managing Borderlands project', UK

Community mapping is participatory mapping that is carried out with groups of community stakeholders or members of a community to identify, communicate, validate and record spatially based data, issues, and locations of importance using paper maps or GIS software. It can also be used to compare options. Forrester and Cinderby (2011) reported on a community mapping event to generate flood management options in Peebles, UK. The first event asked five people representing key local organisations, to mark flood issues on acetate overlayed on maps of different scales. They then highlighted areas where new flood protection or land use changes would help reduce flooding. The information was transferred to a digital GIS format. The possible solutions for local flooding issues were presented at agricultural shows on a map and attendees used flags to mark additional information and comments. This was fed back into the GIS to add another layer of community mapping. It identified areas where different sectors disagreed with solutions (ibid.). See Figure 4.4 in colour plates.

Engagement in water resource planning

An understanding of tools was necessary in order to complete our discussion on developing a community engagement plan. A possible community engagement plan is shown in Table 4.3 and would also be supplemented by time frames, costs, and responsibilities for engagement. It illustrates three time-important activities in engagement:

- the provision of information to stakeholders
- the gathering and aggregating of stakeholder input and dialogue about issues
- providing feedback on how stakeholder input affected the decision.

It also reinforces that information needs to be provided early to stakeholders and throughout the life cycle of the process to claim transparency and build trust. It illustrates that different methods are more suitable at different stages of the planning process. It also illustrates that meeting the minimum standards required by most existing legislation is simply not best practice consultation. Good stakeholder processes inevitably go well beyond legislative requirements.

Tips about consultation

Go where the people are that you are trying to reach:

Table 4.3 Possible community engagement plan for water resource planning

Planning stage	Engagement purpose	Water advisory committee (WAC) role	Other engagement
Initiation	Identify issues Identify stakeholders Inform public and stakeholders of process, implications and opportunities to participate	Review stakeholder analysis and consultation plan and provide feedback	Stakeholder analysis Advertised notice of commencement, proposed timetable and engagement plan Targeted and general invitation to respond on issues, stakeholders, engagement plan, scope, timetable Brochures about planning process, time frame
Situation analysis	Advise on trends, environmental, cultural and social values, opportunities, risks and threats Seek community input, data, knowledge, and feedback	Review draft Situation Report before public release Input local knowledge about resource characteristics, significant natural and infrastructure assets	Targeted consultation by technical teams developing reports using focus groups and interviews Release of background information paper; fact sheets or brochures Input local knowledge about resource
Setting objectives	Identify community aspirations Obtain comment on draft objectives	Discuss objectives	Advice from peak stakeholder groups if necessary or ad hoc group of invited persons covering a range of interests
Identifying and assessing options	Developing socio-economic impact assessment of options Identifying options and trade-off possibilities	Suggest and review strategies and options; preliminary assessment of impacts of options or scenarios Recommendations for strategies and operational mechanisms	Targeted consultation with stakeholder groups using focus groups, interviews, scenario modelling as appropriate

Draft Plan	Test draft plan objectives, strategies, and implementation measures for workability, readability, and acceptability Input on anything that is missing or poorly addressed	Review draft(s); clarify interpretations prior to public release Facilitate discussions with constituencies Members contribute to WAC submission	Public advertising of release of draft plan inviting submissions Available on website Targeted consultation with stakeholder groups Public meetings or workshops
Final Plan	Sharing of issues identified in submissions, meetings and workshops Input on changes as a result of draft plan review and submissions	Review issues against changes in plan WAC makes recommendation to Minister	Targeted consultation with key stakeholder groups only if changes affect or are of interest.
Final Plan endorsed and gazetted	Ensure understanding of plan implications	Briefing of WAC Feedback on process	Public release of Plan Briefings with peak stakeholder groups on implications Fact sheets on background and management implications

(Source: Adapted from Hamstead *et al.* 2008b)

- Involve locals in setting up the process
- Watch the use of language, dress for the occasion
- Train invited speakers to be sensitive to the needs of the audience.

If necessary, provide incentives or other support for participation. This might include a good meal, transport, or a newsletter that participants can take away and share with others. Figure 4.5 (see colour plates) illustrates engaging women and children as the target implementers of a water quality improvement project in Indonesia.

Ensure skilled facilitation to equalise power. A skilled facilitator ensures that everyone involved is fairly heard and not criticised, and that the process is set up to support participation by key stakeholders.

Three other interrelated matters need to be addressed: barriers to consultation; skills needed for effective consultation; and evaluation of consultation. In many cases, an agency is opposed to thorough consultation, possibly due to previous negative experience. This may be a result of poorly designed processes or lack of skills of those undertaking engagement. Without evaluation of the consultation process, it is difficult to justify engagement or know where to target improvement.

Barriers to engagement

Well intentioned water resource planners developing a consultation process are in some cases met with scepticism about the benefits. Some of the arguments used against consultation and possible responses are documented in Table 4.4.

Skills required for consultation, consensus building, and conflict resolution

In many cases, an agency officer who is an expert is hydrology or agricultural development is allocated the job of consulting with stakeholders, frequently with little relevant training or support. Guidance may be in the form of what was done before, even if it was not very effective.

The Hamstead *et al.* (2008a) review of water planning in Australia documented a lack of skills in consultation, consensus building and conflict resolution among agency staff responsible for water planning, who often come from a technical background. Yet to run a good process, at minimum good facilitation, communication and project management skills are needed to supplement the resource and technical knowledge often provided by other team members. Many of the issues that arise in relation to water resource planning do so because of perceptions of fairness in process (procedural equity) or the way resources are distributed (distributional equity).

An effective and neutral facilitator can ensure that participants are heard and their interests are acknowledged. An independent person is more likely

Table 4.4   Responses to arguments against community engagement

Barriers and arguments against community engagement	Response
Politicians are elected to represent the public and know what their constituents want.	Politicians are seldom elected unanimously yet they are supposed to represent and help address the concerns of all their constituents. In many cases, even with the best intent, it is impossible for them to be in touch with the wide diversity of their stakeholders' needs. The process and outcomes can be a litmus test for an approach, a possible lifesaver at election time.
Engagement will delay the process: fear of increased time and costs.	The cost of engagement is a small proportion of the total cost of any planning or development process. A well designed process can be cost-effective and save time overall by reducing delays due to conflict.
Decisions are technical and the public is not qualified. Technical experts know the right solution. Unlikely to use input if community is not highly educated, or believe greater good should prevail.	Community can contribute local knowledge and suggestions to mitigate negative impacts. A good process will ensure that information is presented in an understandable way and investment is provided to enable informed participation, e.g. if literacy is low or access to technology is poor.
Past experience of poor process, possibly token, or seen as 'therapy'.	Lack of skills in facilitation and conflict resolution. Lack of agreement to process.
Management and support of advisory committees is too resource intensive with unhelpful outcomes.	Good facilitation, communication and project management skills can contribute to a good process. A team member allocated to consultation can be a long-term cost-effective solution.
Consultation provides a platform for dissidents and unrepresentative groups, and leads to paralysis and inaction.	Lack of skills in facilitation and conflict resolution. Lack of agreement to process. Ensure techniques are tailored to outcomes sought and diversity of stakeholders.
Government concerned with change; unwillingness of planner, administration, or politician to share power.	Consultative processes requires specific skills to manage well.
Fear of public scrutiny of decision-making process. Lack of transparency; possible insecurity.	Knowledgeable, unbiased and skilled staff can support good policy and decisions.

to test assumptions held by participants, work more effectively across organisations, and have a clearer delineation of roles. Basic public participation practice recommends that the facilitator is selected, approved, and/or acceptable to all parties. In fact, participants should be able to raise any concerns about a facilitator, and to change practitioners if necessary (Sarkissian *et al.* 2003).

In addition, given water's physical and economic characteristics, decisions about water allocation have the potential to generate conflict. Conflict can involve private individuals, farm businesses, water companies, municipalities, industry sectors and can be between provinces, institutions within the same government, and nations with cross-boundary watercourses. So not only are facilitation skills necessary, but capacity in consensus building and conflict resolution is a practical asset.

While water legislation may or may not include provisions for settling disputes (e.g. by agreement or tribunal adjudication), better training and skills in this area can contribute to reduction in serious conflict. Basic understanding of consensus building and conflict resolution should include the following:

- Establishing ground rules to build relationships – A premise of consensus-building is to negotiate as if relationships matter (Fisher *et al.* 1991; Susskind *et al.* 1999). Fostering trust and building relationships is facilitated by good ground rules established early in any consultation process, whether in a public meeting, workshop, or an advisory committee, to ensure all are heard and respected.
- Neutrality – In processes requiring conflict resolution, the requirements are clear: a mediator or 'third party neutral' should not accept his/her role if there is a conflict of interest (actual, perceived, or potential). Neutrality is essential to gain trust and no amount of skills or strategies is likely to encourage resolution if there is a perception of lack of neutrality. As one of the roles of a facilitator is to keep the process on track, all parties need to trust in his/her independence. As with facilitation, all parties should have fair opportunities to participate and be heard. A third party neutral treats parties even-handedly by applying equal interventions, maintaining a balance between parties' verbal exchanges, and focusing equally on each party. The person uses non-judgmental language and asks open-ended questions; is aware of his/her own personal values; encourages parties to exchange opinions; and allows parties to propose options (Charlton and Dewdney 2004). While third party neutrals can often not directly alter any power imbalance between parties, they have some responsibility through their control of the process (Boulle 2001: 225). They can influence and exert power by being present, managing the setting, enforcing ground rules, using certain techniques (such as active listening, reframing, summarising), and if necessary terminate the process (Boyle 2005). When one party has exceptional power, there is little inclination to negotiate or incentive to make trade-offs and find an acceptable solution to all parties.
- Deliberation – The greater the opportunity for deliberation the better chance that a wide variety of views can be incorporated in a decision and the greater the opportunity to ask questions and share values and interests. Good communication can result in mutual understanding whereas uncontrolled debates often prevent people from understanding each other's issues (Hogan 2002).

Joint fact-finding is a collaborative technique for achieving acceptance of data. The concepts of 'joint fact finding', 'co-production of knowledge', and joint problem-solving are made possible by adoption of a consensus building approach whereby knowledge is jointly constructed by stakeholders and experts instead of just reported by experts to stakeholders (Amengual 2006; Ehrmann and Stinson 2006). Effective resource management needs to be based on sound knowledge. Yet ecosystem-human interactions are highly complex, continually changing and characterised by uncertainty, and even more so because of climate change. Cullen (2004: 1) refers to the global knowledge bazaar and the need to obtain the right knowledge at the right time. He argues the need for three clusters of knowledge to interact: local, scientific and Indigenous knowledge, in order to achieve sustainable management.

Other process steps such as identifying a range of options are crucial. While this and the ability of making trade-offs is discussed in greater detail in chapters 7 and 8, a transparent and thorough review of options (i.e. an expanded 'pie') enables a greater range of possible solutions to address diverse needs.

Evaluation of community engagement

Because participation now plays an essential role in water management and infrastructure development, an evaluation of the participation is essential if practitioners are to support their claims about its benefits with clear evidence. Yet it is seldom done. The absence of objective evaluation means a much-reduced ability to learn, manage adaptively, and continually improve practice.

Approaches to evaluating participation generally centre around whether it meets the good principles for engagement mentioned previously. Based on literature on evaluation of participation (Mackenzie *et al.* 2009; Laurian and Shaw 2008; Baldwin and Twyford 2007; Michels and De Graaf 2010; Rowe and Frewer 2000), we propose a framework for evaluating participation that includes assessment criteria categorised as process, output, outcome, and impact indicators. Examples of these types of criteria follow:

Process criteria tend to include:

- the nature and extent of involvement by appropriate stakeholders – was it inclusive, flexible to needs of participants with relevant information shared (i.e. information access)?
- the existence and strength of rules supporting the effective sharing of views – were all key stakeholders able to voice their concerns and join in discussion (i.e. deliberative and fair)?
- the introduction of participation early in the decision-making process;
- the commitment of the agency to the process and its response to public input – was it made clear to participants how their input influenced the decision (i.e. transparent)?
- being cost-effective.

Output criteria (sometimes referred to as 'short-term outcomes') tend to include:

- the extent of agreement on some or all key issues
- adequacy of the information stakeholders can understand and accept as accurate
- the making of feasible proposals.

Outcome criteria, also referred to as second and third-order effects, can be categorised as direct and indirect. They include:

- a plan that serves the interests of all stakeholders
- a plan that is flexible enough to be adapted to new conditions
- the success with which public values are incorporated into decision-making
- the extent to which engagement achieved its purpose
- resolution of conflict
- improved working or personal relationships (e.g. increased trust in public agencies)
- the widespread perception that outcomes are just or serve the public interest.

Impact (or influence) criteria might include:

- the degree to which the public influenced the final decision or outcome
- the extent to which decision-making is delegated
- commitment to implementing the outcome (Baldwin and Twyford 2007).

Two different methods are generally used to assess participation effectiveness:

a) statistics and document analysis – the number of meetings, percentage of identified stakeholders attending; comparing participant suggestions with final outcomes; uptake or compliance with decision; and
b) surveys of participants or external others to gain an impression of satisfaction with the process and impressions of the extent to which it achieved the outcomes and had an impact.

Ideally the evaluation process should be determined at the outset of the process and if possible have independent oversight (Baldwin and Twyford 2007). A sound logic frame can assist. For example, if goals of engagement were to encourage social learning, build trust, change practices, or resolve a particular conflict, indicators and methods need to be appropriately incorporated.

Rigorous assessment of the effectiveness of engagement is a challenge given the limited budgets of most organisations for planning. In addition, the complexity of water resource management leads to challenging and contestable issues, with its wide range of stakeholders; large geographical

---

Box 4.9: Examples of evaluation

An evaluation of regional Natural Resource Management (NRM) groups in Australia found that more guidance is needed to improve outcomes around fairness, including how to deal with potential conflicts between the public and private interests of board members; equitable allocation of resources; and addressing the interests of voiceless future generations. Governing bodies and stakeholder groups need to accept that, in order to engender cooperation, accountability is multi-directional: downward, outward as well as upward (Lockwood *et al.* 2010: 998).

Despite a growing emphasis on participation, at the time the World Commission on Dams report was released in 2000, it was estimated that around 50 per cent of dam projects still did not plan for public participation by people who would be affected. In a global assessment of participation on dams and dam development some years later (UNEP 2007), key issues were identified as:

- not enough time, resources, and information were made available
- the spectrum of participants was narrow; affected people and minority groups often were not included
- participation was limited and was confined to late stages in the process
- real change depended on taking legal action
- affected people were not included in the design and implementation of the process
- staff assigned to monitor the process usually were not trained in public participation.

In some of the cases examined, outcomes were affected adversely by a lack of funding. In order to obtain the commitment of agencies to funding stakeholder participation and implementing the outcomes of that participation, it is essential to demonstrate its positive effects and benefits. Moreover, in order to learn from the experience of a given project and improve subsequent practise, it is necessary to undertake objective, independent evaluation. In none of the cases examined was public acceptance directly and objectively measured (UNEP 2007).

---

coverage; the geographic scale of downstream effects and over-extraction or degradation; and uncertainty in attributing cause and effect given the multiple influences on potential outcomes.

While comprehensive evaluation of public participation per se is ideal, it is seldom undertaken in large projects. Typically, if an assessment is done, it focuses mainly on the process and only occasionally on the outcome, despite the perception of practitioners that process, outputs, outcomes, and impact are closely related (Baldwin and Twyford 2007). Evaluation needs to be accompanied by capacity-building in skills and techniques to achieve effective stakeholder participation so that practitioners and agencies do not feel threatened.

Community engagement in water planning is essential to make effective implementable decisions. Engagement should occur in every aspect of the process for planning but it needs to be tailored to the purpose of process and needs of the stakeholders. With limited fiscal resources and sometimes a sense of political urgency, it is an area where shortcuts are often taken. Yet implementation is easier if stakeholders understand the reason for and have input into decisions. For this reason sound and independent evaluation of plans and the process are important in order to improve both. In each of the next chapters, the role of stakeholders is apparent.

5 Situational analysis

Making good decisions about water resource management begins with a situational analysis to develop an adequate understanding of the water resource, the benefits it provides, risks to those benefits and potential future demands on the system. A historical emphasis on physical attributes and demand assessment has led to an information imbalance, with it now recognised that additional effort is needed to understand social, economic and environmental implications of resource management decisions and to consider issues such as climate change which affect future water supply security (Parris 2011). Such assessments also provide a baseline for future monitoring, evaluation and adaptive management.

This chapter identifies the information needed for a situational analysis. It also discusses ways to undertake assessments when there is insufficient data, time or resources for thorough investigation.

Elements of a situational analysis

To a large extent water resource planning addresses competing benefits, for example between the benefits associated with taking water out for consumptive water use, and the benefits associated with leaving water for non-consumptive uses such as fishing, recreation, tourism, and biodiversity. It aims to ensure that using water for some benefits does not come at the expense of others, or at least if there is a loss of benefits that this is a conscious and transparent choice. Priority is usually given to initiating water resource planning in places where water supply demands are at or approaching the limits of what can be provided, or are adversely impacting on other benefits.

A situational analysis provides baseline information on the nature of the current and projected benefits and uses of the water resource, and how they are related to water resource management. It seeks to answer these questions:

1 What are the benefits currently derived from the water resource, and what new or expanded benefits are planned or desired?
2 How are these benefits related to water resource management?

3 What are the risks to these benefits if water resource management arrangements are left unchanged?
4 Where are their opportunities to improve benefits or reduce risks to benefits through changes to water resource management?

In response to the first question, most commonly, situational analysis is focused on benefits related to water supply for consumptive use and/or hydro-electric power generation. This includes assessments of:

- Current and future demands for consumptive water use for cities and towns, irrigation and industry;
- Current and proposed dams and related infrastructure for water supply, hydroelectricity generation, flood control or other uses that might affect a river's natural flow;
- Likely future inflows to rivers or recharge to aquifers;
- Models of how the water resource behaves, including how available water is currently shared and, where relevant, how infrastructure is currently operated, to assess how well projected water use demands will be met.

Recognising the underpinning benefits of ecosystem health, assessments of water-dependent ecosystems or valued aspects of those ecosystems (e.g. iconic species) are increasingly being done. Less common are assessments of benefits that are not related to extraction of water, for example natural flood mitigation (e.g. through flood peak attenuation via wetlands and meanders); water quality maintenance; visual amenity and aesthetics for mental health and social cohesion; tourism and recreational uses of lakes, wetlands and rivers; and harvesting of fish and vegetation from rivers by riparian populations for food.

One comprehensive way of addressing all of these benefits is to use an ecosystem services framework. The UN Millennium Ecosystem Assessment (MEA 2003: 8) defines an ecosystem as:

> a dynamic complex of plant, animal, and microorganism communities and the non-living environment interacting as a functional unit.

Rivers and aquifers are ecosystems that provide services that are beneficial to humanity. The UN Millennium Ecosystem Assessment characterises these services as:

- provisioning services which include supply of water for towns and agriculture, but can also include direct harvesting of fish and plants, timber grown in wetlands, supply of genetic materials, etc.;
- regulating services including regulation of floods, disposal of wastes, groundwater discharge and recharge, and maintenance of water quality;

- cultural services that provide recreational, aesthetic, tourism and spiritual benefits;
- supporting services which underpin all the above such as biodiversity maintenance, soil formation, photosynthesis, and nutrient cycling.

The second question is aptly addressed by Plant *et al.*'s (2012) framework for applying ecosystem services to water resource planning, which is built around identifying benefits and beneficiaries and linking them back to the water resource via services. The framework differentiates the services provided by ecosystems from the benefits received by people in order to avoid double-counting in economic evaluations and environmental accounting, and to link ecosystem services, benefits and beneficiaries in a transparent way.

Under this framework (illustrated in Figure 5.1):

- Beneficiaries are the people to whom benefits from aquatic ecosystems accrue.
- Benefits are the gains in well-being dependent on ecosystem services that are obtained by beneficiaries.
- Services are the features, characteristics or conditions of the aquatic systems upon which benefits depend. The distinction between a service and a benefit as used here is simply that a benefit typically requires some contribution from people, whereas an ecosystem service requires only the ecosystem.
- Processes are similar to the MEA's 'supporting services'. They are split into two parts. One is the hydrologic processes relevant to water resource planning, labelled as the 'water regime'. These are the characteristics of an

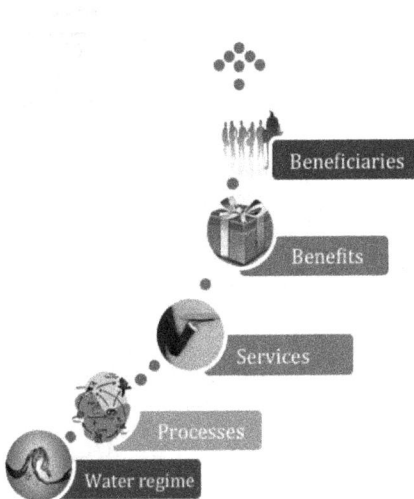

Figure 5.1 Linking water resources to benefits via services
(Source: Plant *et al.* 2012 p. 6. Used with permission.)

aquatic system's water regime (natural or modified) upon which services depend or, put simply, water regime characteristics that support services. The second part is labelled 'other processes, functions or conditions'. These are often greatly affected by factors outside of the scope of water resource planning (e.g. riparian degradation by livestock, water quality degradation arising from catchment land use change). Understanding them though, can facilitate establishing collaborative work to coordinate actions to support benefits.

Plant *et al.* (2012) propose setting out these elements in a table (as shown in Figure 5.2), so that relationships can be more easily understood and analysed. They suggest completing the table by first identifying benefits and beneficiaries using community engagement methods, then using scientific assessments to develop an understanding of how these benefits are dependent on services and processes, in particular the water regime. Plant *et al.* (2012) provide a listing of typical benefits and associated services that are relevant to water resource planning that can be used as a prompter in identifying a full range of benefits and beneficiaries. For convenience this is reproduced in

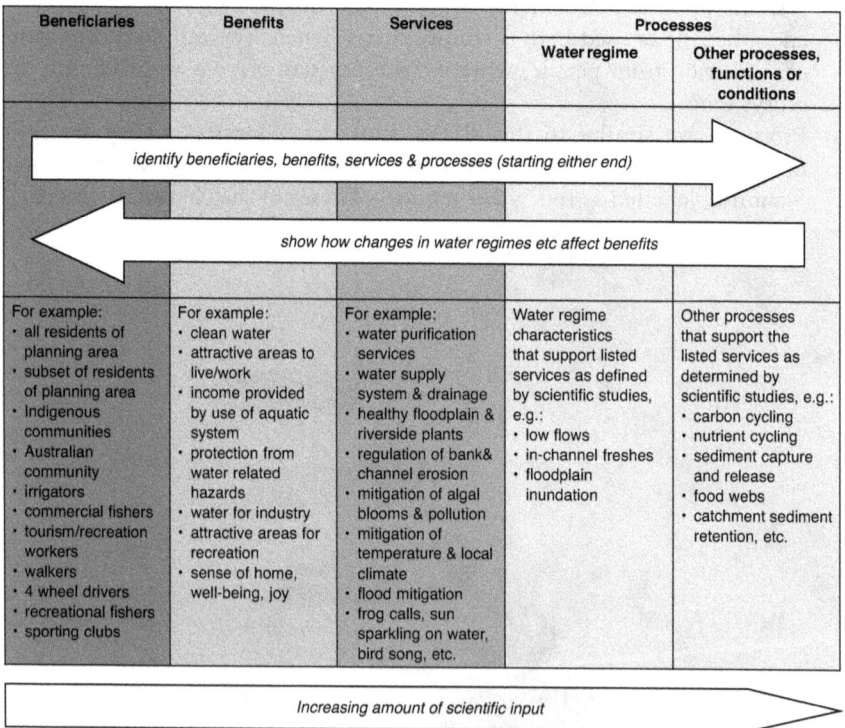

Beneficiaries	Benefits	Services	Processes	
			Water regime	Other processes, functions or conditions
identify beneficiaries, benefits, services & processes (starting either end) →				
← *show how changes in water regimes etc affect benefits*				
For example: • all residents of planning area • subset of residents of planning area • Indigenous communities • Australian community • irrigators • commercial fishers • tourism/recreation workers • walkers • 4 wheel drivers • recreational fishers • sporting clubs	For example: • clean water • attractive areas to live/work • income provided by use of aquatic system • protection from water related hazards • water for industry • attractive areas for recreation • sense of home, well-being, joy	For example: • water purification services • water supply system & drainage • healthy floodplain & riverside plants • regulation of bank& channel erosion • mitigation of algal blooms & pollution • mitigation of temperature & local climate • flood mitigation • frog calls, sun sparkling on water, bird song, etc.	Water regime characteristics that support listed services as defined by scientific studies, e.g.: • low flows • in-channel freshes • floodplain inundation	Other processes that support the listed services as determined by scientific studies, e.g.: • carbon cycling • nutrient cycling • sediment capture and release • food webs • catchment sediment retention, etc.
Increasing amount of scientific input →				

Figure 5.2   Examples of elements of an ecosystem services framework

Source: Plant *et al.* 2012 p. 12. Used with permission.)

an Appendix at the end of this book. Different kinds of assessments to define the relationship between benefits and the water regime are discussed later in this chapter.

Ecosystem services can apply at a local, regional, national and global scale. Biodiversity maintenance for example is flagged as being a global issue in the UN Convention on Biodiversity. This Convention recognises that the capacity of the earth's ecosystem as a whole to continue to supply services that support humanity and to recover from natural (and man-made) disasters is reliant on biodiversity. Consideration of biodiversity is reflected in policies and legislation in many countries. Thus one cannot solely rely on community engagement to identify all benefits and beneficiaries – some supporting services such as biodiversity are not so readily linked to vocal or even visible local beneficiaries. Laws and policies often fill this gap by mandating action to preserve threatened or endangered species or habitats. Regardless, consideration of potential effects on important habitats and species is normally an important part of situational analysis. Applying an ecosystem services approach like this ensures that planning considers a much broader range of benefits that may be affected by water management, rather than simply those associated with water supply.

The approach to gathering this information typically involves two phases. The first is an initial assessment that collates the information currently available. The second is the filling of information gaps to the extent determined to be necessary for planning to proceed. This can lead to a two-phased approach to water resource planning, where the initial assessment leads to some interim actions, with more comprehensive actions being determined after information gaps are filled.

An example of this is the water resource planning process applied for the Padthaway groundwater area in South Australia. Initial water resource planning in 2001 identified that water tables were declining and groundwater salinity was rising such that, if no action was taken, water use benefits for irrigation would be lost. However at that time there was insufficient information to determine why salinity was rising, and what rate of water extraction would be sustainable over the long term. In December 2001 the Minister for Water Resources declared a Notice of Restriction on the taking of water, to hold water extraction at current levels while sustainable management options were developed. At the same time studies were instigated to address the knowledge gaps about groundwater hydrology and quality and a community process commenced to develop management options. Once the studies were completed a plan was made in 2009 that included actions to address the identified risks (see SENRMB 2011).

Identifying benefits and beneficiaries for a water resource can be conveniently done in association with a stakeholder analysis at the start of any planning process (see Chapter 4). It should provide information to populate the first two columns in Figure 5.2. Given this information, the next stage is to develop an understanding of the services and processes that underpin these benefits, and the risks to them under current management arrangements.

There are many variations in terminology and approach to risk assessment, but fundamentally they all share the common conceptualisation of risk as a combination of the likelihood of a threat to something of value, and the consequence of the occurrence of that threat. In water resource planning, benefits described above are what is of value, and the water-related threats to them can include such things as water shortages and water quality deterioration, either on a short term event basis or on a long-term trending basis. For risk assessment it is important to understand the relationship between these benefits and the water resource, and their vulnerability to changes in it.

In the remainder of this chapter we describe several common types of assessments that are done to provide this kind of information. These assessments should include consideration of risks associated to the benefits under current management arrangements, and opportunities for changes in resource management to increase benefits and reduce risks, as per the third and fourth questions listed at the start of this chapter.

Water resource modelling

Understanding how water behaves in a water resource underpins many other assessments. A water resource model is a commonly used method to document information in a way that can be used during situational analysis as well as later in the planning process.

Water resources are a defined part of the water cycle in an area. A water resource model delineates the boundaries of the resource and elements within it to the level of detail needed for the planning process. It defines where and how water moves in and out of the system and where and how it is stored and moves within the system.

Water resource behaviour can be quantified in a model in terms of a static (average or sample) or dynamic water balance based on the following equation:

Inflow – outflow = change in storage

What makes up each of the three elements of the equation depends on the boundaries and features of the water resource, and the level of complexity and detail that is possible and desirable. Inflows to and outflows from the resource can be broken down spatially, for example by location, zone or river reach; by type, for example extraction for human use, evaporation, outflow to the sea, etc.; and temporally, showing how they vary over time. Additionally the behaviour of water within the resource can be described to varying levels of detail showing for example flows of water from one area within the resource to another.

A *conceptual model* showing the key elements of the water resource and hydrologic relationships is the essential first point. Figures 5.3 and 5.4 show simple, schematic, conceptual models for two different kinds of water resource. They show the resource as a whole, addressing inflows, outflows and storage. They could be expanded to show movement within the resource, different areas of internal storage, and different locations and kinds of inflow and outflow.

Figures 5.5 and 5.6 (see colour plates) show examples of where a conceptual model is populated with quantified data. In these 'static water balance' examples quantified information is added that in total specifies a water balance for the resource, in the case of Figure 5.5, for one year, and for Figure 5.6, as an estimated long-term average.

Internationally, there is movement towards standardising the way such information is described and presented. The System of Environmental-Economic Accounts for Water (United Nations Statistics Division, 2012) provides the conceptual framework for water account preparation within countries and on an international scale. In addition to physical water balance parameters, it covers economic uses of water and related monetary information. It includes agreed concepts, definitions, classifications, accounting rules and tables, and data items and recommendations on the methods to compile them. Many countries are in the process of implementing water accounting based on this framework.

In addition to water balance information, statistical information on variability of the water regime set out in the conceptual model over time can be compiled to provide a greater understanding of resource behaviour. Where suitable historic records are available, statistical information on trends and events such as drought and floods, identifying frequency, duration, extent, etc., can be compiled in tabular and graphical form. As discussed below,

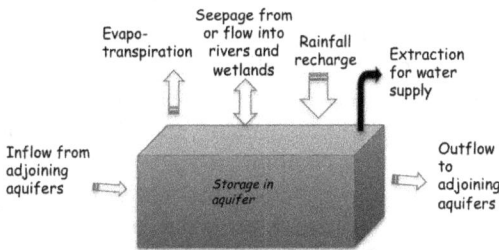

Figure 5.3 Simple groundwater resource conceptual model

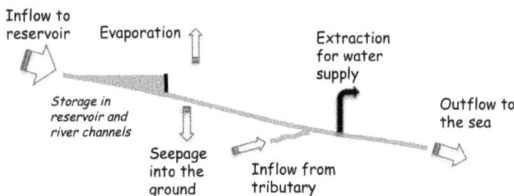

Figure 5.4 Simple river water resource conceptual model

projections for the future can be based simply on past records, but preferably consideration of future uncertainty should lead to a range of scenarios.

Which statistics are relevant depends on the identified benefits, as discussed later in this chapter. For example information on drought sequences and frequencies is likely to be very important for assessing risk to water supply benefits; information on floods for flood benefits; information on flow parameters for fish migration events for a fishing industry; and information on wetland flooding for ecosystem benefits.

Where information and capacity is available and it is warranted, more complex dynamic models can be built. Dynamic models simulate the movement of water in, out and within a water resource over time. They involve representing the physical characteristics of the resource, using mathematical relationships and measured or estimated physical parameters such as storage and channel capacities, aquifer transmissivity, etc. The water resource is broken into connected components, or elements, which are internally homogeneous and the model simulates the inflow, outflow and storage of water in each element and the flow between elements over time.

Often river inflow information is not available so it is generated from rainfall data using rainfall-runoff models. These models use catchment characteristics and other parameters to generate river flow data from rainfall data.

A large variety of dynamic water resource modelling tools are available for rivers, aquifers, catchments, water distribution systems and combinations of these. Selection and development of dynamic water resource models requires specialist skills and knowledge that are not covered in this book but are well addressed elsewhere (see for example Barma and Varley 2012, Loucks and van Beek 2005, Barnett *et al.* 2012).

Selection of the appropriate modelling tool should be determined based on selecting the simplest tool to achieve what is needed, i.e. representing the adjustable management arrangements and the identified risks to and opportunities for benefits. A common output is the extent that water is available to meet water supply demands, but additionally outputs relevant to environmental and other non-consumptive benefits should be available.

The more physical processes the model simulates, the greater its need for data and the greater the cost, expertise and effort needed to construct and use it. For example, three-dimensional models that simulate river levels as well as flows, storage and extraction are much more complex and need much more data than two dimensional river models that do not deal with water levels. If river levels are not needed for planning, then two-dimensional models are a much less costly solution.

For services and benefits in addition to water supply, the relevant flow regime aspects need to be determined through additional assessments. Examples of where more complicated dynamic modelling need to be supplemented by environmental, economic or social studies include: flow rates and levels for flooding needed to maintain the condition of ephemeral wetlands; water levels at specific locations for significant trees or wetlands to access groundwater; high flow event timings for fish breeding and migration; minimum flows at points to meet water needs for recreation such as river rafting; or minimum waterhole levels for hippo survival (see further discussion later in this chapter regarding environmental water assessments). Whatever these parameters are, the model selected should be able to generate the outputs needed to assess the extent to which the services and benefits can be influenced by water resource management.

Dynamic water resource models need to integrate information on current management rules including:

- how decisions are made for day-to-day operation of infrastructure such as dams and weirs and diversion works. This can include obligations to meet water levels for navigation, channel capacity constraints, flood control, etc.;
- obligations to supply water, for example where there are legal or contractual requirements to supply water from dams to meet town water supply, hydropower, recreational water levels, environmental needs, or water licences or permits;
- priorities and methods for sharing water when demand exceeds availability;
- any rules and methods that protect water from being extracted for purposes such as ecosystem health or water quality maintenance.

Such information is normally readily available from government agencies and infrastructure operators.

In addition to understanding the conceptual basis for models, it is important for water resource planners as well as key stakeholders to appreciate the uncertainties in model results. Water resource models, of necessity, are always a simplification of a complex reality. While parameters such as aquifer transmissivity, channel roughness and resource boundary conditions can be estimated by physical observations at points, they cannot be measured at all points, and so models use representative parameters. The representative parameters are often determined by a process of model calibration, which in simple terms means applying known inputs (e.g. rainfall, river inflows, water levels, extraction quantities) and adjusting the unknown parameters until the model as closely as possible simulates a known output (e.g. river outflows, storage levels, water levels). For example in a river system model, data from gauged inflows and metered water extractions over a period of time might be used as model inputs, and various internal model parameters such as channel roughness are adjusted until the simulated flow at a point closely represents the measured flow that occurred during the same time period. Validation of the calibration can be done by calibrating the model parameters using part of the available time series of data, then validating the calibration by testing how well the model simulates the remaining available time series of data. Confidence in the calibration is increased where the calibration and validation periods cover a wide range of the possible inflow and extraction sequences.

Where calibration is limited or not possible due to a lack of measured data, a sensitivity analysis is often used to estimate the uncertainty in the model. This involves varying model parameters around the estimated parameter value within a range that is considered to be feasible to assess the extent to which model outputs are affected by the change. Results give one indication of the uncertainty in the model results.

Future scenarios for the water resource

Because water resource plans are forward looking, a prediction of future rainfall, inflow or recharge is a requirement. When combined with water resource models this enables future available water to be estimated. The most common approach used up until recently has been to assume future rainfall and river flows would be similar to available historic records of rainfall and river flows (typically from the previous twenty to one hundred years). For static models this means using long-term statistical figures derived from historic information, such as annual average, median, 90 percentile dry, or driest on record.

More recently there has been greater consideration of future climate scenarios outside of the range of recorded historic data, to reflect evidence of climate change or longer term climatic cycles beyond the period of recorded historic rainfall and river flow records. Scenarios are different from forecasts: scenarios explore a range of possible outcomes resulting from uncertainty; whereas forecasts identify the most likely pathway and estimate uncertainties. Therefore, 'scenario planning is not forecasting of the most probable future but it creates a set of the plausible futures' (Amer *et al.* 2013: 25).

Climate change offers an additional level of uncertainty, with the IPCC concluding that 'observational records and climate projections provide abundant evidence that freshwater resources are vulnerable and have the potential to be strongly impacted by climate change, with wide ranging consequences' (IPCC 2008). Lower rainfall in places will likely mean more sunshine, higher temperatures, and therefore higher evapo-transpiration rates. At a time of increased irrigation demand from storages and rivers, it also leads to less surface runoff to rivers at the same time as the environment is also water-deprived. Faurès *et al.* (2010) report that small variations of rainfall usually translate into much larger variations in river runoff. They referred to an example of analysis of historical precipitation and runoff data in Cyprus that showed that an average reduction of 13% of annual precipitation translated into a 34% reduction in runoff. Likewise in the Murray-Darling Basin in Australia, a 13% decline in rainfall observed in the southern MDB from 1997–2006 (which included drought years) resulted in a 44% decline in streamflow (Horne 2013: 142). In terms of modelling future flows in the Ebro basin in Spain, Garcia-Vera (2013: 226) reports that:

> For the 2010–2040 timeframe, a 12% average decline for the Ebro basin was estimated, or what is equivalent to a 17% decrease in surface flows,

9% in throughflow, and 13% in underground runoffs. Impacts will be greater in the summer months and problems aggravated in the dry months.

This can have devastating effects on providing a reliable supply particularly in arid and semi-arid areas. Furthermore, less runoff combined with higher temperatures can affect river water quality, e.g. through increases in blue-green algae.

Recent work by Australia's Commonwealth Scientific and Industrial Research Organisation (CSIRO) (see Chiew *et al.* 2008) considered a range of future inflow scenarios for water resources in the Murray–Darling Basin as follows:

- a continuation of the climate of the past 112 years
- a continuation of the climate of the recent 10 years (very dry)
- projections of climate in 2030 taking account of global warming.

Global warming scenarios were determined using data from a set of 15 internationally recognised global climate models, which simulate the effect of rising levels of greenhouse gases in the atmosphere on regional rainfall patterns. Data from three different emission scenarios (low, medium and high emissions) for each model were used, giving in total 45 data sets. From these scenarios, three were selected as representing upper, lower and median levels for possible average future inflows. This approach was intended to encompass uncertainty in both future emission levels and in the models themselves. Derived future climatic statistics were then imposed on 112 years of historic climatic data to produce adjusted datasets for the purpose of assessing risks and comparing management scenarios.

The CSIRO then combined these climatic scenarios with water resource models to simulate resource behaviour under different development scenarios to provide information on associated flows and availability of water for extraction and watering of environmental assets. See Box 5.1 for an example of how this information was used in the Northern Victorian part of the Murray Darling Basin.

Non-flow data can also be used to improve understanding of possible future flows by extending knowledge about flows in the past, thereby capturing more information about long-term climatic cycles. For example, the historic flow record for the Oldman River in Alberta Canada was extended back 628 years using tree ring data, revealing greater inter annual and inter decadal variability than the data from river gauges over the last 110 years, including more severe drought years in the early 1700s and longer periods of low flow in the 1840s to 1870s (Corkal *et al.* 2011).

Scenario development is not without challenges. Models are only as reliable as the data and need to be validated. Decisions need to be made about the time and spatial scales for scenario development – between 5 to 20 years, a river reach or entire basin. Scenarios can be hard to justify and require effort in developing explanatory material and graphics to convince stakeholders of their authenticity. For scenarios to be credible and seriously considered by participants they need to be: internally consistent; plausible; relevant and useful; creative (challenge the prevailing mindset to consider options beyond the status quo and organisational comfort zone); and transparent (Amer *et al.* 2013). In Chapter 4 we referred to the effectiveness of visual methods for engaging stakeholders and scenarios frequently lend themselves to being portrayed using visual techniques.

A valuable benefit of scenarios is in providing the ability to deal with uncertainty without making a specific commitment.

Box 5.1: Northern Victoria, Australia, Sustainable Water Strategy

In the midst of a record-breaking drought, the government of Victoria in Australia developed a new water allocation strategy for its northern region. In doing so estimates of possible future climate scenarios were made. Figure 5.7 shows estimates of trending changes in inflows compared to average historic that were used to produce adjusted future climate scenarios.

Forecast availability of inflows for the total Murray system* over 50 years
(Scenarios A to D compared with the long-term average)

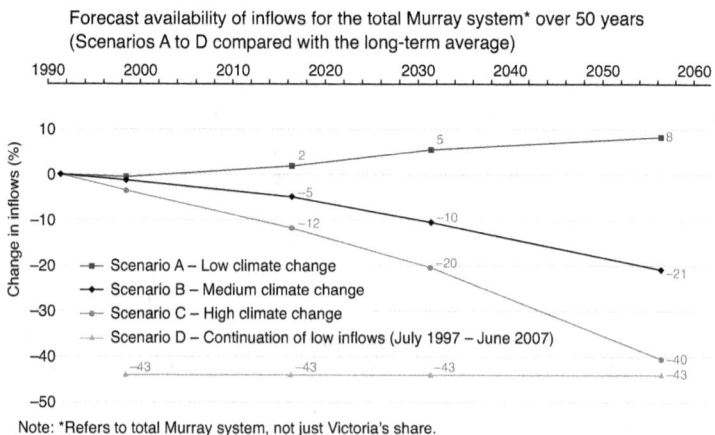

Note: *Refers to total Murray system, not just Victoria's share.

Figure 5.7   Future inflow scenarios used in developing the Northern Victoria
              Sustainable Water Strategy

(Source: Victorian Government 2009: 22. Used with permission.)

Using system models, estimates of the effect on water available under water entitlements were made for different future climate scenarios. An example of a graphical representation of the different long-term water availability is shown in the supply-duration curves in Figure 5.8. The 'base case' is climate from the past 112 years repeated.

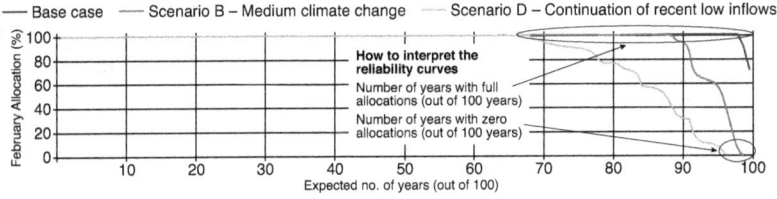

Figure 5.8   Future water availability scenarios used in developing the Northern Victoria Sustainable Water Strategy

(Source: Victorian Government 2008b: 55. Used with permission.)

Models were also able to produce data for particular parameters considered important for environmental health. Figure 5.9 shows for each year the number of months when average river flow was below a threshold level for river health.

Forecast numbers of months where recommended flows of 10 ML/days are not met in the Campaspe River (base case and Scenarios B and D)

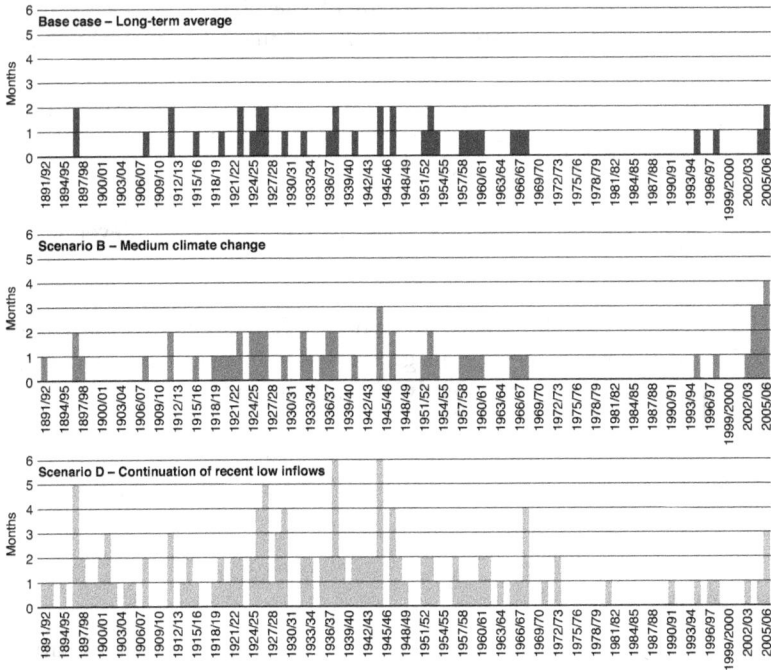

Figure 5.9   River health indicator under different future climate scenarios used in developing the Northern Victoria Sustainable Water Strategy

(Source: Victorian Government 2008b: 59. Used with permission. For more information refer to Victorian Government 2008b: 50–59)

Groundwater – surface water connectivity

There is increasing recognition of the fact that groundwater and surface water are connected parts of the water cycle, and that altering or taking water from one is likely to have an effect on the other. Depending on the nature of the connection, the taking of groundwater can have a one-to-one impact on river flows within a short period of time, or can have less than a one-to-one effect that is time lagged by years or decades. The former often happens in alluvial aquifers that adjoin rivers, while the latter can apply to any case where the connection is less strong and direct, for example deeper porous rock aquifers where the effect is delayed by intervening less porous strata.

In the past, separate water resource plans have commonly been prepared for groundwater and for surface water resources. Flows across connections are represented by fixed data sets (e.g. river 'losses'). The risk is that the effect of management decisions for one water resource on the other water resource may not be considered properly. Coordinated decision-making across the two water resources is most warranted if the linkage between resources is high.

SKM (2011) proposes that connections be characterised on the basis of:

- Potential for connection – assessment of the physical surface water and geological environments and the way they connect, based on the character of the aquifer (e.g. deep alluvial, shallow alluvial, sand, fractured rock, porous rock, etc.) and the rivers (perennial or ephemeral, whether regulated by infrastructure or not) and the nature of the connection (contiguous or noncontiguous, gaining or losing);
- A time lag between extraction and impact ('time to impact');
- Qualitative influences such as climatic conditions and level of water resource development.

The information from this characterisation provides a guide as to the vulnerability of river flows to change in groundwater management (and to some extent vice versa) and consequently the nature and extent of coordinated management likely to be needed.

Other approaches to characterising and assessing surface to groundwater connectivity are discussed elsewhere (see Winter *et al.* 1998, Brodie *et al.* 2007).

Water supply benefits

Water supply assessments can provide a picture of the level, and nature, of dependence of the community on the use of water taken from the resource; its economic value; and future trends in demand for water. It includes water for cities and towns, irrigation, mines and industries. Future demands for urban water can be estimated from current water use, population projections, and already planned technological, regulatory or economic measures to restrain consumption. Urban water utilities are usually responsible for providing such information. Irrigation, industry and mining demands can be estimated from current demand plus planned or likely expansion or contraction, drawn from available information on new or proposed developments and government development plans. Future demands for irrigation can be estimated based on projections of crops, areas and application rates. Box 5.2 illustrates water demand assessments prepared in Arizona.

In some areas, allocation will depend on the amount of water available from a particular source rather than the demand or request alone. The assessment will then need to create an understanding of the benefits that are potentially forgone by restraining access to water, or alternatively the cost of meeting the shortfall from other water sources.

Economic valuation techniques are readily applied to assessment of the value of benefits accruing from water supply, and the cost of possible supply shortfalls under current management arrangements. The value of irrigated agriculture can be assessed from information on crop types, production quantities and extent of crop areas; crop economic value both at the farm gate and regionally; comparison with value of non-irrigated crops in the area; direct and indirect employment related to irrigation; and local/regional related expenditure. The cost of shortfalls will depend on the available options. For example, where this is an option, economic value can also be assessed by determining the cost of supplying the water from alternative water sources – e.g. recycled urban water, groundwater, desalination plants, or long distance pipelines. For irrigated agriculture the cost may be the difference between irrigated and dryland crop returns, or the cost of purchasing and shipping in fodder for livestock.

Box 5.2: Water demand assessments in Arizona, USA

Faced with declining groundwater levels, rapidly expanding urban areas, and a need to demonstrate effective water management to justify Federal assistance with development of the Central Arizona Project (the CAP) to divert water from the Colorado River to developing areas of the state, the government of the state of Arizona in 1980 legislated for twenty five years of planned water management, consisting of five cycles of planning and implementation at approximately ten year intervals. In preparation for the fourth water management plan the Department of Water Resources in 2006 conducted a detailed assessment of past and projected water supply and demand for each area of managed ground-water resource (designated as 'Active Management Areas' or AMAs).

Data on water taken was derived from annual returns submitted to the Department by water users from 1985 to 2006. Post-2006 projections were prepared by analysing a range of data and selecting three baseline scenarios to represent a likely range of possibilities: Scenario One: lowest reasonable water demand: Scenario Two: demand in-between Scenario One and Scenario Three; and Scenario Three: highest reasonable water demand.

Generally, the difference in municipal demand between the three baseline scenarios is due to a combination of assumptions regarding future population growth and corresponding water use. The difference in agricultural demand in the three baseline scenarios involves different assumptions concerning whether irrigable lands will be fully farmed, and whether certain irrigated lands will be taken out of production for residential development. For Indian agricultural demand, it was assumed that by 2025, the amount of irrigation on the reservation would increase, with different assumptions on the rate of increase in each scenario. The primary difference in industrial demand figures concerns assumptions regarding population growth and electrical power generation.

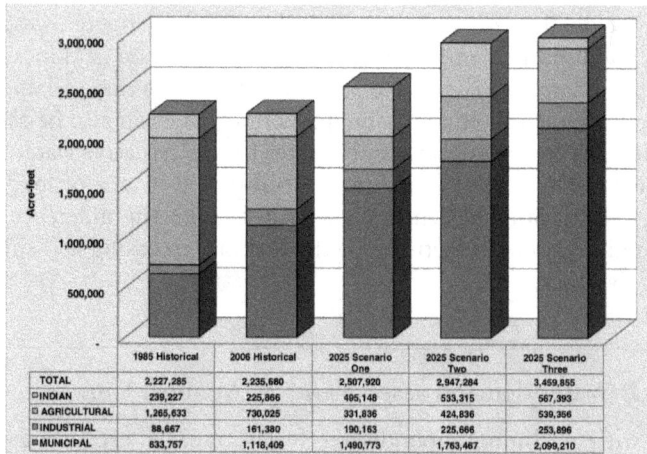

	1985 Historical	2006 Historical	2025 Scenario One	2025 Scenario Two	2025 Scenario Three
TOTAL	2,227,285	2,235,680	2,507,920	2,947,284	3,459,855
INDIAN	239,227	225,866	495,148	533,315	567,393
AGRICULTURAL	1,265,633	730,025	331,836	424,836	539,366
INDUSTRIAL	88,667	161,380	190,163	225,666	253,896
MUNICIPAL	833,757	1,118,409	1,490,773	1,763,467	2,099,210

Figure 5.10 Water demand projections, Phoenix Active Management Area (Source AZDWR 2010: 58. Used with permission.)

The assessment also overlayed a range of scenarios for sources of water to meet these demands, including groundwater, local river water, CAP water, and recycled urban water.

Risks to these benefits can be assessed using information on the extent to which the water needed to meet water supply demands is likely to be met under forecast future climate scenarios with current management arrangements. Where projections of demand exceed projections of water availability from resource modelling, shortfalls can be identified, whether these are rare, periodic or ongoing. This gives an indication of the likelihood of shortfalls in supply. This can be coupled with information on the cost of water supply shortfalls to determine an overall risk in economic terms under current arrangements.

Where risks are identified, the assessment should consider whether there is an opportunity, by altering water resource management arrangements, to address the risk. One approach is to consider whether the water shortfalls are 'absolute' or 'economic', where an economic shortfall represents the case where there is sufficient water but current infrastructure is not sufficient to enable it to be fully utilised. A study by the International Water Management Institute in 1998 estimated that globally just over a billion people live in arid regions facing absolute water scarcity by 2025, and 348 million people face severe economic water scarcity by 2025 (Seckler *et al.* 1999). Additionally, opportunities to address shortfalls through altering access limitations or infrastructure operations to make more water available in times of shortfall should be identified (noting that these usually come at the expense of either long-term average water available for water supply, or water for other purposes such as the environment).

For water resources where there is potential and demand for expansion in water supply, the situational analysis can give an indication of the opportunities in terms of locations, quantities, required infrastructure, etc. However this would only be indicative, scoping the possibilities, with detailed assessment of different levels of expansion normally occurring later in the planning process (see chapter 8).

Many of the social and community benefits of water supply for towns and irrigated agriculture production are not so readily quantifiable. The Hamstead *et al.* (2008) review of water planning in Australia found that socioeconomic studies were frequently not undertaken, but when they were, they usually focused on economic rather than social implications, varied in quality, were 'non-aligned' or relied on secondary data (Baldwin *et al.* 2009). This may be because water resource planners often rely on a participation process to alert them to major social issues. However, credible studies that provide an accurate profile early on in a process can offer more robust and objective evidence and input to decision-making. The social profile mentioned in Chapter 4 provides a good basis for identifying gaps and the need for further information. In fact, useful guidelines for objective social and economic assessment of water resource benefits (IACSEA 1998) were prepared for use in water resource planning in NSW, Australia but were seldom implemented.

Non-economic indicators of the benefits that come from water supply can also be derived based on information about communities and their dependence on water, derived from surveys and other statistical information. Community sensitivity or resilience indices have been developed and used extensively in water resource management. For example in preparing a water resource plan for the Burnett River in Queensland, Australia, the government commissioned a study that assessed the community sensitivity to changes in water availability for irrigation (see Hausler and Fenton 2000). This study collated and analysed data on towns in the area from national census records to estimate how sensitive the community would be to changes in water use businesses (largely irrigation farms) arising from changes in water availability. Two types of indices were prepared and plotted as shown in Figure 5.11.

The first index combined census information on age, education and occupation, unemployment, income, family and housing, to give an overall indicator of the communities' ability to adapt to change in any kind of locally employing business. The second index combined data on the number of water use businesses and numbers of occupied dwellings within each area to give an indication of the dependence of the area on water use businesses. Plotting the two indices together gives an overall indication of sensitivity to change.

The information on community sensitivity to water availability can be coupled with data on periods of future shortfalls in supply to give qualitative information on which areas can benefit (or be impacted) most by changes to management arrangements.

Non-consumptive benefits

As discussed earlier, there are a wide range of benefits that can be derived from water resources that do not require water to be extracted from the resource. These include economic benefits such as hydro-electricity generation and tourism, through to social and cultural benefits such as recreation, social relations and cohesion, amenity and cultural heritage and identity.

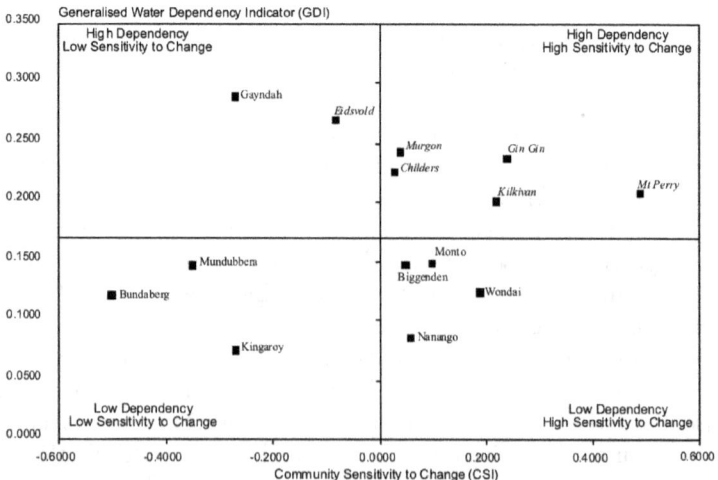

Figure 5.11 Community sensitivity to water availability index, Burnett River, Queensland, Australia

(Source: Hausler and Fenton 2000: 69; © Department of Natural Resources and Mines, Queensland, Australia. Used with permission.)

Information on the economic benefits of such things as power generation, better water quality and tourism can be readily obtained through economic valuation techniques including business turnover, value, and costs, as well as employment generated. Water requirements are usually easily determined, e.g. availability of water at the turbines for power generation, or minimum (and maximum) water levels in rivers and storages to allow tourist operators and tourists to boat, raft, fish or to access particular sites such as wetlands. Water resource models can be designed to estimate the extent that these water requirements are likely to be met in the future. This can be coupled with economic tools to estimate the value of lost output that is likely to occur under current arrangements. Flood mitigation benefits can also be valued using methods such as cost to repair flood damage. The assessments can also likewise consider shortfalls related to projected new developments, e.g. proposed hydroelectric infrastructure, expanding cities or towns.

Opportunities to address risks to those benefits can be identified by assessing whether there is potential to change water resource management arrangements to address the identified shortfalls. It is important to differentiate between those cases where the water resource plan can make a difference and those where it cannot (e.g. large floods, periodically drying rivers with no prospects for infrastructure) so the plan can be focused on where it can make a difference.

Often more challenging is determining the water requirements for benefits that are not so readily defined in economic terms, such as community benefits from aesthetics, cultural heritage values, subsistence fishing, spiritual aspects. These tend to be more qualitative in nature, with greater use of community surveys and other such techniques to estimate water requirements and what happens when there is a shortfall in water.

The example in Box 5.3 illustrates use of social and economic profiling to better understand community dependence on water resources in the River Murray.

Box 5.3: Examples of social and economic profiling, River Murray, Australia

Because of the need to reduce water extraction, the River Murray, at the lower end of the Murray-Darling system in south Australia, has been the subject of a few studies to identify social and economic values related to water.

Study one – social and economic profile
The 'Strategic Impact Assessment of Possible Increased Environmental Flow Allocations to the River Murray System' (Hassall *et al.* 2003) presented social and economic profiles for each catchment in the study

area against a number of descriptors or variables, identifying the dependence of communities on the water resources and ecosystems.

Two features of this study are noteworthy:

- Not only did it include analysis of national statistical and other data, but also featured primary data collection, i.e. over 90 interviews were used to corroborate and add to findings.
- A broad range of descriptors were used, including, among others:
  - assessment of dependence on water by irrigated agriculture as well as non-agricultural industries such as tourism, recreation and fishing
  - community well-being and services
  - cultural heritage (sites of Aboriginal, non-Aboriginal and natural heritage)
  - cumulative change and community response (interviews regarding change over last 10 years and how community responded).

Areas were ranked according to which were least dependent on water or least affected by the cumulative impacts of recent changes, and those which were somewhat dependent and most dependent on water (Baldwin *et al.* 2009).

Study two – Nested use of 'Concept' and DMCE

In 2011, Mooney *et al.* (2012) ran a series of workshops exploring social values in environmental prioritisation in the context of water scarcity in a section of the River Murray in South Australia. They used a simulation modelling tool, 'Concept', to identify what aspects of the wetland environment the community values; the social use of wetlands; and the interaction of the two value sets and where they overlap. Three stakeholder groupings – a water manager group, a water user group, and a representative group comprised of various interests including conservation, Indigenous peoples, youth, rural women, etc., developed different conceptual models.

A wide suite of use and non-use values was identified. A deliberative multi-criteria evaluation (DMCE) was run with the representative group and scenarios demonstrated how changes in the weighting of different values would impact on the allocation of environmental water. The most important value identified was ecological health, followed by social well-being and then, threatened species and habitat. Other identified values were significant sites, recreation, economic production, and research and education. The information on values contributed to a better quality of deliberation based on a shift away from entrenched values and understanding of others' interests (Mooney *et al.* 2012).

Many Indigenous populations around the world are disadvantaged and pose a challenge about how to address their values in water resource planning. In Australia, the National Water Initiative recognises Indigenous interests in water and requires jurisdictions to include Indigenous customary, social and spiritual objectives in water resource plans. In spite of this, water resource plans rarely specifically address Indigenous requirements (NWC 2009; 2011c). Partly this is due to an assumption that environmental flows will serve as a surrogate to meet these requirements. Partly it is due to the difficultly in trying to determine how to include their interests in a statutory plan. In addition, there are 'substantial legal and other structural impediments to increased Indigenous access to valuable water rights, particularly for commercial purposes' and 'little guidance is provided to water resource managers ... relating to Indigenous access and involvement' (Jackson *et al.* 2012: 2). However a first step with any Indigenous people is to try to understand their relationship and linkages with water and ensure they are considered in the water planning process.

Some tools have proved useful to elicit qualitative data on Indigenous values. Mooney and Tan (2012) worked under an informal research protocol on the River Murray, with the Nganguraku who describe themselves as 'River People'. Methods involved a guided site visit with traditional owners, the use of photo documentation and a focus group meeting. The choice of methods was based on the preferences of the traditional owners as well as pragmatic considerations of available time and resources. The focus was on identifying places, or things of value that were affected by the quantity and quality of water in the resource, rather than to quantitatively define the relationship between Indigenous values and flow regime. The outcomes were used to inform the water resource planning process (Jackson *et al.* 2012).

Similarly, a suite of methods was used to identify issues of concern and increase knowledge of Indigenous issues in relation to Condamine Alluvium (groundwater) water resource planning. These included semi-structured interviews, visits to country, photographic recording and mapping of places of significance, and a desktop review of published cultural heritage studies. Traditional owners identified deterioration of important sites for medicine and food plants, dried springs and fish kills as a result of hydrological changes of stream flow and aquifer depletion from over-extraction for irrigated agriculture. Their ability to pass on traditional knowledge and fulfil custodial responsibilities to country were also impaired (Tan *et al.* 2012).

In the Ord river basin in Western Australia, a specific study was commissioned to establish, record and articulate the cultural values that Indigenous traditional owners and communities attach to the Ord River (Barber and Rumley 2003). Traditional owners with knowledge of and interests in the area were engaged during the fieldwork period, and various locations along the Ord River were visited with these people. They were able to discuss, articulate and record their associated cultural, social and economic values, including religious beliefs and environmental knowledge as indicated in the relationship and interactions between themselves and the Ord River ecosystems. Discussions included reference to ecosystems within the Ord River valley both in the pre- and post-dam contexts. Site protection and minimising or avoiding impacts on Indigenous cultural values within the study area were also discussed, as were approaches by which these might be achieved (DoW 2006). Because the waterway has already been modified, some of the cultural values have changed or been lost. Consequently, the study acknowledged lost or changed values and indicated Indigenous interest in contributing to future waterway management and related policies and plans.

Indigenous aspirations for water use and management in Australia have been summarised as:

- to have their spiritual relationships to water resources protected
- to protect sites of significance
- to decisively participate in the better management of water resource
- to exercise their spiritual, cultural, social and economic rights through access to good quality water and water for commercial use (Jackson and Morrison 2007).

Ecosystem assets and processes

As mentioned earlier, the UN Millennium Ecosystem Assessment (MEA 2003) defines an ecosystem as 'a dynamic complex of plant, animal, and microorganism communities and the non-living environment interacting as a functional unit'. Particular units of ecosystem called 'ecological assets' (e.g. a particular river or wetland) and particular 'processes' within ecosystems (e.g. water regime, carbon cycling, etc. – refer to Figure 5.2) that support services and benefits can be identified.

For situational analysis, assessments can be done to identify 'important' or more valued water related ecosystem assets and processes, and assess their condition and the water related risk to their maintenance. Value can be assessed in terms of benefits.

It is recognised that this overlaps with the previous discussion regarding assessing water supply and non-extractive benefits, since ecosystem assets and processes support these benefits. The examples in Box 5.3 illustrate this. In its recently published classification system for ecosystem goods and services, the United States EPA points out the risk of 'double counting' if both ecosystems and ecosystem functions, and the benefits to humans derived from them, are delineated separately (Landers and Nahlik 2013). However we argue that relying solely on the other assessments is likely to result in missing out of important values that are broader and more long-term in nature such as biodiversity, and that it does not matter if assessments overlap provided this is recognised in the planning processes that follow.

Identifying important ecosystem assets and processes

Most surface water-dependent ecological assets are readily identifiable: rivers, wetlands, estuaries and floodplains. Identifying groundwater-dependent ecological assets and understanding their water requirements is an area of fairly recent development. Groundwater-dependent ecosystems include rivers (in relation to baseflows arising from groundwater), in-aquifer and cave ecosystems, groundwater-dependent wetlands, groundwater-dependent terrestrial vegetation, estuarine and near-shore marine ecosystems.

A thorough benefits and services analysis can lead to identifying which ecosystem assets and processes are most important in terms of supporting benefits. For example, in Victoria, Australia, regional government authorities charged with investing in improving the health of rivers undertook an assessment of all the rivers in their area on a reach by reach basis to assess their 'social and economic value' (see, for example, West Gippsland Catchment Management Authority 2005) including:

- use for irrigation, towns or power generation
- use for recreational fishing and boating
- use for camping, swimming and other recreational activities
- presence of a site that has cultural or heritage value.

Additionally, the reaches were assessed for 'environmental value' in relation to a range of criteria including:

- presence of threatened, endangered or vulnerable species as listed under state regulations
- state of vegetation and species as compared to reference levels
- importance for native fish migration
- inclusion of 'significant' wetland, either Ramsar-listed or identified in a national, state or regional listing of important wetlands
- 'naturalness' or how close the river is to its untouched natural state
- 'rarity' in relation to kinds of habitats
- water quality.

Sources of information included field assessments, surveys and government databases. Parameters were scored (or rated) so that the value of river reaches could be compared and higher value reaches identified.

As shown in this example, some services that support broader benefits can be missed where immediate and local benefits only are identified (the 'social and economic values'). Ecosystem assets and processes also underpin community benefits that apply at a national or global scale on a long-term basis. The need to maintain our ecosystem as a whole to support such things as the atmosphere, and the ability of future generations to enjoy a full range of ecosystem services is manifested in global biodiversity conservation agreements, threatened and endangered species protection, and habitat and ecological function preservation policies and laws. In the above Victorian example, these broader benefits are implicitly recognised through the 'environmental value' criteria. While it is important these are not overlooked, this needs to be moderated by cost and availability of data, and whether certain features can provide a proxy for others. For example, while aquatic biodiversity is not explicitly stated in the above list, naturalness and water quality provide a proxy indication.

Sometimes what is important is defined by government policy. For example, the 'Water Framework' Directive (WFD) of the European Union aims to achieve 'good ecological status' (i.e. the protection and enhancement of health and biodiversity of the aquatic ecosystem) in most European rivers, lakes and wetlands by 2015 (European Commission, 2000), and nations commit to maintain wetlands they list under the Ramsar Convention.

Assessing water regime requirements and risks

Assessment of the water regime requirements (flows, water levels, etc.) for maintaining these identified ecosystem assets and functions is critical to inform the water resource planning process. For ecosystems that are in poor condition or in a state of decline, an understanding of cause and effect, often complex, can lead to identification of what it would require for recovery or to stem the decline. For ecosystems in good condition it should lead to understanding what is required to maintain that condition. This can be combined with information on the likely range of water regime characteristics that are projected to indicate where there is or could be a shortfall, and consequently a risk to these assets and functions.

For rivers, these water regime requirements in rivers have commonly been called 'environmental flows' where:

> Environmental flows describe the quantity, timing and quality of water flows required to sustain fresh water and estuarine ecosystems and the human livelihoods and well-being that depend upon these ecosystems (Brisbane Declaration 2007).

There is now wide recognition that environmental flows are more than just minimum flows, but are rather a dynamic, variable water regime that preserves important aspects of the natural flow regime, including such things as high flow events and seasonal variability (Arthington *et al.*2010). As we are discussing both rivers and groundwater, we use the more generic term 'water regime' to cover the temporally and spatially varying flow regime for rivers and the water level/pressure and flow regime for groundwater.

Estimating the water regime needed to maintain a water ecosystem in a desired condition is a scientific assessment that draws on the best available knowledge of causal links between different aspects of the regime and the response of the ecosystem in terms of the services it provides. For rivers, the linkage between water regime and biodiversity has been described by Bunn and Arthington (2002) in terms of four principles:

1 major flow events drive geomorphic processes, determine the form of habitat available for plants and animals and therefore set limits to those that can continue to exist in a river system;
2 the plants and animals associated with a river have evolved responses that match the opportunities and pressures provided by the natural flow;
3 many animal species in rivers, and plant and animal species adjacent to rivers, require the periodic connections that occur during floods if they are to survive, disperse and prosper; and
4 changes to the flow regime reduce the competitive advantage of endemic species and encourage colonisation by exotic and introduced species – and possibly also dominance of a sub set of the existing range of species.

Fundamentally, changes to the flow regime will increase the likelihood of change to the biological and other elements of the river ecosystem. A range of assessment methods has been developed to provide more specific information on the relative importance of particular aspects of the flow regime. Such assessments assume that a certain amount of modification to the flow regime can occur without incurring excessive risk of loss of valued services. Tharme (2003) identified over 200 methods from 44 countries. The methods vary greatly with respect to their objectives, complexity, and the data, technical and scientific inputs they require.

Some methods focus on preservation of particular species, typically fish. These methods focus on minimum flows for maintaining habitat and for spawning and migration. Others are more holistic, and aim to address the support of the ecosystem as a whole.

The 'FLOWS' methodology, developed in 2002 by the Victorian government in Australia, uses a 'bottom-up' 'building blocks' approach (see DNRE 2002). It identifies flow dependent environmental assets, identifies flow component characteristics (standard component types are used) which are required to maintain or enhance them, and assesses the extent to which those characteristics are currently present. Using hydraulic and hydrologic models, the current river flow characteristics can be compared to the required characteristics to assess where there are gaps. For each assessed river reach, the extent to which the current flow component meets the recommended parameters is expressed on a six point scale, ranging from 'mostly' to 'never'.

A 'top-down' approach to environmental flow assessment, such as the Benchmarking Methodology (see Brizga 2007; Quinlan *et al.* 2004), uses the natural or existing flow regime as a starting point, and examines the consequences of altering or removing various components of the flow regime. This compares with the bottom-up FLOWS approach, which identifies minimum flow characteristics needed to support important values. Both of these holistic approaches require field work, information on flows and technical and scientific input. The South African Downstream Response to Imposed Flow Transformation (DRIFT) is also a top-down approach (see King *et al.* 2003). Some literature suggests a combination of a 'bottom-up – top-down' approach would be most useful (Arthington *et al.* 1998, Brizga 2007).

Tidal estuaries and coastal areas require a different form of assessment again. The issue is not so much water levels and rates of flow as ingress of saline water from the sea. The sensitivity of the estuary salinity to changes in dry season flows depends on the hydraulic nature of the system (volumes of water, depth, level of connection to the sea). Sensitivity to flow regime indicators can be derived based on general physical characteristics (e.g. see NSW Office of Water 2011, Appendix 4). Hydraulic models are needed to predict salinity changes with any degree of confidence, with linkages to surveys of habitat and information on habitat response to salinity.

In terms of groundwater, the dependency of ecosystems on groundwater is based on one or more of four regime attributes:

- flow or flux – the rate and volume of supply of groundwater
- level – for unconfined aquifers, the depth below surface of the water table
- pressure – for confined aquifers, the potentiometric head of the aquifer
- quality – the chemical quality of groundwater expressed in terms of pH, salinity and/or other potential constituents, including nutrients and contaminants (SKM 2001).

SKM (2001) proposes a framework for assessing the water regime requirements, consisting of four steps:

1 identification of potential groundwater-dependent ecosystems
2 analysis of ecosystem dependency on groundwater
3 assessment of water regime in which dependency operates
4 water requirement determination.

They also review a range of tools for addressing each of these steps.

Methods such as those discussed above are well documented and can be applied, with some adjustments, to cases in any location. However detailed river-by-river or aquifer-by-aquifer assessments such as these require a level of expense that is often not available and, where the number of rivers required to be assessed is large, will exceed the ability of available scientists. For such cases, lower cost more rapid alternatives are needed, with the option of applying more intensive approaches to particular rivers where the risks are greatest.

Some low cost rapid assessments use hydrologic flow statistics to make recommendations (e.g. protecting flows below the 60th percentile in dry months). These rely on 'rule of thumb' relationships between flow parameters and instream habitat and function recommended by experts based on their experience. In 1999 the state government of New South Wales in Australia adopted a set of River Flow Objectives, which identify descriptively particular aspects of the water regime that were considered to be most relevant for meeting ecosystem needs in NSW rivers (see Box 5.4). These were then interpreted and used as the water regime requirements for ecological assets and functions for water resource plans spanning a large number of smaller rivers where case-by-case assessments could not be resourced.

Box 5.4: River Flow Objectives – New South Wales, Australia

1 Protect natural water levels in pools of creeks and rivers and wetlands during periods of no flow
2 Protect natural low flows
3 Protect or restore a proportion of moderate flows, 'freshes' and high flows
4 Maintain or restore the natural inundation patterns and distribution of floodwaters supporting natural wetland and floodplain ecosystems
5 Mimic the natural frequency, duration and seasonal nature of drying periods in naturally temporary waterways
6 Maintain or mimic natural flow variability in all rivers
7 Maintain rates of rise and fall of river heights within natural bounds
8 Maintain groundwater within natural levels, and variability, critical to surface flows or ecosystems
9 Minimise the impact of instream structures
10 Minimise downstream water quality impacts of storage releases
11 Ensure river flow management provides for contingencies
12 Maintain or rehabilitate estuarine processes and habitats.

The purpose of the river flow objectives is to produce specific environmental benefits such as:

- improved survival of ecosystems and aquatic biodiversity
- improved water quality
- healthier wetlands
- improved habitat quality and increased variability of habitat for native fish, frogs, waterbirds and other native fauna, including invertebrates
- more successful breeding of native birds, fish and other native fauna, which only breed in response to specific environmental triggers, for example, rising or falling water levels in the natural seasons
- more natural inundation of flood plains and wetlands, leading

to better health and productivity (such as grazing), protection of endangered species, biodiversity and water quality
- discouragement of alien pest species, such as carp, which favour regulated conditions
- improved health of instream and riparian vegetation, leading to greater bank stability, improved efficiency of buffer strips in protecting water quality, and reduced erosion and turbidity
- reduced frequency of algal blooms.

(source: http://www.water.nsw.gov.au/Water-management/Water-sharing-plans/Environmental-rules/Rivers/Rivers/default.aspx, as at 3 July 2013)

Assuming that reductions in low flows by extraction of water would increase risk to ecosystems as per the generic requirements in River Flow Objectives 1 and 2 (see Box 5.4), the NSW government prepared regional maps of 'hydrologic stress' for smaller rivers, using a desktop assessment technique, where stress was calculated as the ratio of extraction (based on estimated peak daily demand) during times of low flow (based on a flow that is available for a percentage of time, generally the 80th percentile) (see NSW Office of Water 2011). This information was overlaid onto maps that showed the value of rivers based on an aggregation of information on biodiversity, recreation, tourism, heritage, cultural and other such services. The result was a map of risk to instream values, as shown in Figure 5.12 (see colour plates).

In 2010, a group of internationally recognised environmental flow scientists recommended a methodology that 'bridg[ed] the gap between the simplistic and often arbitrary hydrologic "rules of thumb" presently being used for regional-scale estimation of environmental flow needs and, at the other extreme, the detailed and often expensive environmental flow assessments being applied on a river-by-river basis'. Called ecological limits of hydrologic alteration (ELOHA), it aims to identify ecosystem water regime requirements for many rivers across a large area in a single process. The ELOHA methodology broadly consists of:

- using hydrologic modelling to prepare baseline (i.e. undeveloped) and current hydrographs for all of the rivers for a common time period, preferably at a daily time step;
- classifying river segments into distinctive flow regime types, based on ecologically relevant flow parameters, and sub-classifying them based on geomorphology;
- for each river segment determining the deviation between current and baseline for the relevant flow parameters;

- developing flow deviation – ecological response relationships for each river type using a combination of existing scientific studies, expert knowledge and representative field studies, and thereby estimating the current threat to ecosystems and what would be needed to reduce that threat;
- recognising that there will always be a level of uncertainty and therefore applying an adaptive monitor – evaluate – revise approach to refine understanding of ecological responses over time (Poff *et al.* 2010).

It should be noted that estimates of water requirements for ecosystems will always have a level of uncertainty. Assessments provide values of parameters that are estimated using available knowledge to support the ecosystem. For this reason, water requirements may best be expressed as a range representing high to low likelihood of damage or loss, rather than a single value. Decisions made later in the planning process can then take into account this likelihood of damage in weighing up alternative management options.

Risk and vulnerability

As discussed earlier, risk can be considered a combination of the likelihood of a threat and the consequence of that threat to benefits. In the above discussion of assessments, threats to water resource benefits have been discussed in terms of times when identified water requirements for the benefits are not met. We have also discussed estimating the consequences of the threat in terms of the economic value of benefits lost or reduced, where such valuation is possible. For cases where economic valuation is less straightforward (e.g. important natural assets) valuing can be based on non-economic criteria that prioritise assets in terms of such things as biodiversity, naturalness, as well as any non-consumptive economic benefits. Risk is assessed by combining the likelihood of a damaging event (e.g. a severe water shortage) occurring with an assessment of consequences.

Risk assessments can be improved by adding in assessments of vulnerability to the threat. For example, a town may be very vulnerable if the water resource is its only source of water, but less so if it has alternative supplies available. Similarly, a groundwater-dependent ecosystem can be more vulnerable to a water shortfall if this has been a rare or non-occurring event in the past than if it is already adapted to periods without water. Likewise, a community may be more vulnerable to flooding if it has no evacuation plans or individuals have no insurance.

In NSW Australia a technique using river geomorphology characteristics was used to classify and map river reaches ecosystems that are more and less susceptible to being impacted by human activity (see Hamstead 2010, Cook and Schneider 2006). This 'fragility' classification was defined as the susceptibility/sensitivity of certain geomorphic categories to physically adjust/change when subjected to degradation or certain threatening activities. Three categories were derived:

- Low fragility: resilient ('unbreakable'). Minimal or no adjustment potential. Only minor changes occur, such as bedform alteration, and the category or subcategory never changes to another one, regardless of the level of damaging impact.
- Medium fragility: local adjustment potential. It may adjust over short sections within the vicinity of the threatening process. Major character changes can occur or the category or subcategory can change to another – but only when a high threshold of damaging impact is exceeded, e.g. it may require a catastrophic flood, sediment slug or clearing of all vegetation from bed, banks and floodplain.
- High fragility: significant adjustment potential. Sensitive. It may alter/degrade dramatically and over long reaches. Major character changes can occur or the category or subcategory can change to another one when a low threshold of damaging impact is exceeded.

Threshold points can also be important in assessing vulnerability. There is increasing evidence that ecosystems seldom respond to gradual change in a gradual way. Lakes often appear to be unaffected by increased nutrient concentrations until a critical threshold is passed and the water shifts abruptly from clear to turbid. Submerged plants suddenly disappear and animal and plant diversity is reduced – an undesired state from both a biological and economic point of view. Substantially lower nutrient levels than those at which the system collapsed are required to restore the system. The economic and social intervention involved in a restoration is complex and expensive, and sometimes irreversible. Studies of rangelands, forests and oceans also show that human-induced loss of resilience can make an ecosystem more vulnerable to random events like storms, droughts or fires with which the system could previously cope. An ecosystem with low resilience can often seem to be unaffected and continue to generate resources and ecosystem services until a disturbance causes it to exceed a critical threshold. Even a minor disturbance can cause a shift to a less desirable state that is difficult, expensive, or even impossible to reverse (Swedish Environmental Advisory Council 2009).

There has been a considerable amount of work over recent years to identify ecological thresholds, including such things as critical flooding frequencies for wetland survival and for fish and water bird breeding. Saline intrusion or structural compaction of aquifers are further examples.

Thresholds also apply to human activity. Urban water supply systems often have critical dry sequences, which if exceeded, would result in system failure. Walker *et al.* (2009) reported on a resilience assessment of the Goulburn-Broken catchment in Victoria, which identified a range of known or likely thresholds in the catchment relating to biodiversity, agricultural activity, the local economy, and social value structures.

Trends in condition are important. If condition is declining, it is more likely that an event will trigger a threshold change. On a river system, if connected wetlands or instream drought refuges are reduced in number and area, then it may take only a single event to trigger a major loss of biodiversity. This may even be totally unrelated to the water regime; for example, the event could be a highly destructive wildfire or pollution event.

In summary, assessments of vulnerability and thresholds of change can greatly improve the assessment of risks. A basic prerequisite for water resource planning is having sufficient understanding of the biophysical condition of the water resource and the ecosystem, its use, and the social, cultural and economic relationship of the community with the water resource. Modelling historic patterns that affect river flow and aquifer recharge and using climate forecasts can provide potential scenarios to use as a basis for planning. An understanding of risks to achieving benefits and vulnerability can assist in assessing management options, discussed in Chapter 7. The situational analysis starts to reveal priority issues and is a basis for setting objectives in a plan.

6   Objectives and logic

Given the situational analysis, the *objectives and logic* step in water resource planning is where decisions are made on what the plan should achieve and broadly how it should do so. As discussed in chapter 3, we suggest a hierarchy consisting of non-quantified objectives and outputs (outputs being in effect a lower level of objective), with associated quantifiable performance indicators and targets. We define *objectives* as the desired benefits and associated services that derive from the water resource. We define *outputs* as the water regime characteristics that are expected to deliver those benefits and associated services, subject to key external assumptions. We define *actions* as the means that the plan puts in place to achieve those water regime characteristics. Performance indicators and associated targets are used to indicate quantitatively the extent that objectives and outputs are expected to be achieved and thus the level of trade-off between competing objectives adopted in the plan after consideration of different management options.

Objectives are the foundation of the water resource planning process, guiding strategy development and achieving early agreement on common ground, or foreshadowing the need for trade-offs. Objectives need to be referred to throughout the process to keep priorities in perspective, and may need to be revised depending on additional information uncovered along the way.

In this chapter we look at processes for setting objectives and outputs with their performance indicators. Actions are developed in the next step of planning (chapters 7 and 8). Targets, which indicate the extent that objectives and outputs are to be achieved, are defined once the actions to be adopted are determined, and final monitoring and evaluation mechanisms are set subsequently. We examine these in later chapters.

Purpose for setting objectives

Water resource plans are made to achieve certain objectives, whether those objectives are clearly stated or not. Sometimes objectives are not stated in the plan itself but are implicit in the reason for a plan being made, e.g. a government decision to build a dam for irrigation or for hydropower, where a plan is commissioned to determine optimal dam capacity, or how a dam can be best operated to maximise economic and other benefits. In other cases statements of objectives in plans are vague and general, thereby potentially creating false expectations as stakeholders read into them more than is actually to be delivered.

We argue that a planning process is more effective if the objectives are explicitly and clearly stated in the early stages. Expressly understanding and stating the objectives provides a transparent foundation for developing management options and comparing them in relation to the extent they contribute to or impact on all of the objectives. For example, the taking of water for the objective of irrigated agriculture may result in reduced achievement of other objectives relating to water quality and ecosystem condition. If the criteria for comparing options are based around the level of achievement of objectives it ensures that they are all properly considered.

Taking time to identify and consider objectives of stakeholders can also focus on early agreement on common ground, where objectives are not in competition and there are opportunities to achieve multiple objectives concurrently. For example, preventing saline intrusion into groundwater systems can provide both economic and environmental benefits. It can also help participants to understand the full range of benefits that are being sought. This can be continually referred to throughout the process to assist parties to gain perspective and keep moving forward.

Where assessments of options is done based on the extent they affect objectives, once a decision is made these assessments provide all stakeholders with an understanding of the extent that their objectives are to be met or not. This provides for transparency in decision-making and allows stakeholders to make social and economic choices that are based on realistic expectations of future availability of benefits from a water system.

Well-expressed objectives, outputs and associated performance indicators also form the basis for a meaningful monitoring and evaluation program, enabling the success of the plan to be measured and evaluated over time (see chapter 9). Achievement of the objectives and outputs becomes the yardstick for measuring success or failure.

Processes for determining objectives

A starting point for identifying potential objectives can be broader contextual policy or legislation. For example, the internationally held goals of Integrated Water Resource Management are (Lenton and Muller 2009: 7):

1 economic efficiency – to make scarce water resources go as far as possible and to allocate water strategically to different economic sectors and uses;
2 social equity – to ensure equitable access to water and to the benefits from water use, between women and men, rich people and poor, across different social and economic groups both within and across countries;

3 environmental sustainability – to protect the water resources base and related aquatic ecosystems and more broadly to help address global environmental issues.

These goals reflect the commonly used 'triple bottom line' approach used in assessing impacts or programs. They provide a good basis for ensuring that objectives are comprehensive. A typical water resource plan would be aiming to achieve outcomes in all three of these areas. If there are no stated objectives in one area, it may be that they are being considered but not being explicitly stated. For example, a plan might be developed with a single stated objective of supplying water for irrigation. However in developing the plan the impact on riparian domestic water users and instream ecosystems might also be considered. These implicit objectives should be explicitly stated and discussed up front.

Within a particular jurisdiction there are often objectives in legislation, policies or broader plans that should be considered in setting water resource plan objectives. Boxes 6.1 and 6.2 show how the scope and nature of objectives in water resource plans in the state of New South Wales (NSW), Australia and EU WFD are dictated by contextual requirements.

Box 6.1: Contextual requirements for objectives in water resource plans – New South Wales, Australia

As part of signing up to Australia's National Water Initiative Agreement in 2004, NSW agreed that its water resource plans would include objectives that:

1 define environmental and other public benefit outcomes[1] (clause 37)
2 define resource security outcomes (clauses 37, 43).

Under the State Government's *Water Management Act 2000*, water resource plans prepared under the Act are to 'promote' the following (see sections 5 and 9):

a) water sources, floodplains and dependent ecosystems (including groundwater and wetlands) should be protected and restored and, where possible, land should not be degraded; and
b) habitats, animals and plants that benefit from water or are potentially

---

1 The NWI defines public benefit of water use and management as 'mitigating pollution, public health, for example, limiting noxious algal blooms, Indigenous and cultural values, fisheries, tourism, navigation and amenity values'.

affected by managed activities should be protected and (in the case of habitats) restored; and

c) the water quality of all water sources should be protected and, wherever possible, enhanced; and

d) the cumulative impacts of water management licences and approvals and other activities on water sources and their dependent ecosystems, should be considered and minimised; and

e) geographical and other features of Indigenous significance should be protected; and

f) geographical and other features of major cultural, heritage or spiritual significance should be protected; and

g) the social and economic benefits to the community should be maximised.

Additionally, water resource plans are to be assessed as to the extent they materially contribute to the achievement of the relevant State-wide natural resource management standards and targets in the relevant catchment management area (section 43A).

Box 6.2: Contextual requirements for environmental objectives, EU Water Framework Directive (WFD)

The environmental objectives are defined in Article 4 – the core article – of the WFD. The aim is long-term sustainable water management based on a high level of protection of the aquatic environment. Article 4.1 defines the WFD general objective to be achieved in all surface and groundwater bodies, i.e. good status by 2015, and introduces the principle of preventing any further deterioration of status. A number of exemptions to the general objectives that allow for less stringent objectives, extension of the deadline beyond 2015, or the implementation of new projects, provided a set of conditions are fulfilled.

A Guidance Document on exemptions to the environmental objectives clarifies exemptions such as for technical feasibility, disproportionate cost, alternative means. Disproportionate cost, for example, is a 'political judgement informed by economic information and an analysis of the costs and benefits of measures' (EC 2009: 13). The assessment of costs and benefits has to include qualitative costs and benefits as well as quantitative. The document states that 'the margin by which costs exceed benefits should be appreciable and have a high level of confidence' and should take place after examination of the most cost-effective solutions (EC 2009: 13).

Contextual objectives are typically broad and generic in nature. While it is possible to simply paste them directly into a water resource plan, we argue that to achieve the previously stated purposes for having objectives in a water resource plan, it is important for objectives to be tailored down to be more focused and locally relevant, for example by referring to specific local environmental assets, towns, irrigation districts, communities, etc. A situational analysis assessment as described in Chapter 5 can provide the basis for doing this.

A stakeholder analysis that identifies water system benefits and beneficiaries will point to potential objectives for the water resource plan in a way that is locally relevant. The potential objectives are simply the maintenance or achievement of each of those benefits. Applying an ecosystem services approach in the stakeholder analysis as discussed in Chapter 5 can ensure that the full range of potential economic, social and environmental benefits, and consequently objectives, are identified.

Second, the assessments of the water requirements of identified benefits, and the current state of and risk to those water requirements (also described in Chapter 5), can lead to an indication of the extent to which the water resource plan might contribute to the achievement of the objectives.

Given this information, potential objectives can be identified, sorted and prioritised according to criteria such as the following:

- objectives that reflect broader contextual requirements, including any required topics and minimum conditions;
- the extent of potential benefits associated with the objective, or the risk if the objective was not to be achieved;
- connected objectives, i.e. where achievement of one objective contributes to another;
- the extent that factors outside of the scope of the plan affect the achievement of the objective, and the risk that those factors will govern regardless;
- objectives reflecting benefits most at risk from the continuation of current water management arrangements;
- the extent that the water resource plan could contribute to the achievement of the objective.

To be of value in the planning process, objectives should be comprehensive of all significant benefits that can be affected by the plan, but at the same time be focused on benefits where the plan is most likely to make a substantial difference. Ranking of objectives based on the above criteria can lead to a clearer understanding about what aspects the plan should focus on. Decisions can be made to drop some lower-ranking potential objectives in favour of a clearer focus on objectives where the most can be gained.

It is also important to have a level of flexibility, allowing objectives to be altered as the planning process proceeds and further information and insights are revealed.

Processes for determining outputs

At the same time as objectives are being determined, potential outputs to achieve them can be identified. For a water resource plan these represent the water regime characteristics that are needed to support achievement of the objectives. These are typically such things as low flows, flooding frequencies, reliability levels for water available for extraction or for hydro-electricity generation, groundwater levels, etc.

Assessments of water requirements in the situational analysis step will point to the kinds of water regime characteristics that are relevant. If a benefits table is developed in situational analysis along the lines of that proposed by Plant *et al.* (2012) (refer to Figure 5.2 in Chapter 5), these characteristics will already be identified. The better the knowledge of cause and effect relationships between aspects of the water regime and ecosystem services and benefits, the more precisely these outputs can be defined.

Expressing objectives and outputs meaningfully

Once objectives and outputs are identified, they need to be articulated in a way that is meaningful for the planning process and for stakeholders by making them specific and locally relevant, again using information gleaned from the situational analysis.

Hamstead (2009: 21) illustrates the importance of this with respect to a water resource plan for a river that is supposed to support the Macquarie Marshes, a Ramsar-listed wetland. Supplying water to the Macquarie Marshes is a major feature of the plan, yet the objectives of the plan make no mention of it. The only related objective is the very general:

> maintain or enhance the ecological functions and values of riverine environments (NSW Government 2003: section 10).

He observes that:

> Lack of specificity undoubtedly causes confusion. The Macquarie Marshes as a whole is an area of some 200,000 hectares, of which approximately 20,000 hectares are the Ramsar listed wetland. It is not clear what area and what state of health the water sharing plan is targeting. Commentators reporting that only 10 per cent of the Macquarie Marshes is in good health imply that the target is the whole Macquarie Marsh area, but the reality is that the consumptive use of water for irrigation inevitably means that there will be some loss.

Plant *et al.* (2012) propose that objectives and outputs can be effectively expressed using information from an ecosystem services assessment done during situational analysis along the lines of that shown in Figure 5.2. The objectives can articulate clearly the benefits and beneficiaries and the services that support them, and the outputs can articulate the water regime character-istics that the plan should aim to provide to support he objectives, as shown in Table 6.1.

As is shown in the above examples, objectives and outputs for a water resource plan relate to what the plan aims to achieve and not to the process of plan making. For example objectives such as:

- identify and understand threats to water availability and quality
- engage with the community to develop practical management options
- have a transparent decision-making process

are relevant to the process for making the plan, rather than what the plan itself aims to achieve, and should be clearly distinguished and should not be included within the plan (though they may be included in documents that explain the plan making process).

Finally, an important method to test proposed objectives is to engage with stakeholders to assess whether they are understandable, sufficiently compre-hensive and relevant. This is one of the reasons for continuing stakeholder engagement throughout the process.

Populating the logic framework

In the process of developing the objectives and outputs the logic framework for the plan can be commenced to provide a basis for further plan devel-opment. This includes identifying likely evaluation questions and performance indicators.

The objectives and outputs should have corresponding performance indicators and targets. Performance indicators allow achievement of the objectives and outputs to be measured and evaluated. Targets defined in terms of the performance indicators provide specificity about the extent that objectives and targets are expected to be achieved.

Table 6.1   Detailed articulation of objectives and outputs

OBJECTIVE			OUTPUT
Beneficiaries	*Benefits*	*Services*	*Supporting water regime*
Attract people to the region and support residents'...	...physical and mental health, by increasing amenity and lifestyle opportunities such as walking/ exercising, barbecuing, boating and swimming...	...by retaining attractive native riverside vegetation and fit-for-purpose water quality.	Providing seasonal low flows, bank full flows and inundation of areas shown on map P.

OBJECTIVE			OUTPUT
Beneficiaries	*Benefits*	*Services*	*Supporting water regime*
Support commercial and recreational fishers, businesses associated with recreation and tourism, and Indigenous communities...	...by maintaining and improving fish catches...	...through maintaining habitat and flows for fish breeding and growth in the areas shown on map x	Protect low flows and freshes, and other specific components of the flow regime that support fish breeding and growth.
Support floodplain graziers...	...by promoting fodder production...	...by providing floodplain water and associated sediment in the areas shown on map x.	Facilitate annual inundation of at least the areas on map x for a minimum duration of 2 days.
Support Indigenous communities, visitors and ecotourism businesses...	...by safeguarding cultural heritage values associated... with river pools, Kings Wetland and groundwater-dependent ecosystems x and y...	...through maintaining aquatic habitat, riparian vegetation and water quality.	Maintain groundwater and pool levels within a range adequate to maintain aquatic habitat and riparian vegetation.
Support irrigators...	...by maintaining current levels of irrigated agriculture...	...through providing adequate reliability and quality water.	Protect flow regime components important in water purification services and in providing adequately reliable water supply, as well as services supplying and draining water.
Provide residents of Palatia, Windsor and King's Cliff...	...with water for drinking, commercial use and gardens consistent with current town growth projections...	...by providing reliable water of fit-for-purpose quality at relatively low treatment costs.	Protect flow regime components important in water purification services in providing very high reliability water (Palatia, Windsor), and maintain the position of the aquifer's saltwater interface within its natural range (Kings Cliff).
Support local, national and international communities as well as future generations...	...by contributing to genetic resources and biodiversity, and future options...	...by increasing overall condition of floodplain and aquatic flora and fauna.	Provide water regime characteristics that improve floodplain and aquatic flora and fauna in priority areas and maintain it in the majority of other areas.

(Adapted from Plant *et al.* 2012: 35. Used with permission.)

Selection of performance indicators would commence at this step of planning alongside defining objectives and outputs. They would be used and refined during the next step where management options are compared. The performance indicators should be used to compare those different options in relation to their contribution to achieving outputs and objectives. Once a management option to be implemented is determined, the expected levels of the performance indicators become the targets for the plan.

A summary of the information that should be prepared by this step is shown in Table 6.2, using the logic framework from Table 3.3. For convenience, design of performance indicators is discussed further in the context of monitoring and evaluation in chapter 9. The next chapter examines the range of management options that can be considered for use in meeting objectives.

Table 6.2   Population of logic framework after setting objectives step

Level	Performance indicators and targets	Monitoring and evaluation	Key assumptions
Objectives Objectives should be stated. They should cover all the areas where the plan is expected to have a substantial effect.	Proposed indicators for the objectives and outputs should be selected, including the nature of likely targets that will quantify the objectives and outputs. Target values to be finalised later in the planning process.	Evaluation questions should be established, and practical means for monitoring identified to ensure that performance indicators can be measured. Final monitoring arrangements can be developed later.	Assumed actions and influences outside of the scope of the plan, upon which the achievement of the objectives and outputs are also reliant, should be documented along with development of outputs and performance indicators.
Outputs Outputs that will contribute to the objectives should be identified, and the expected cause-effect relationship documented.			
Actions These are considered in the next step of planning.			

Figure 1.1   Small water harvesting dam ('sand dam') in Tana River catchment, Kenya
(Photo courtesy of C Baldwin 2012)

Figure 1.2   Sustainable rice intensification, near Lake Victoria, Kenya – the system increases
productivity with fewer inputs (water, seed fertiliser) and more widely spaced
plantings (Uphoff 2003)
(Photo courtesy of C Baldwin 2012)

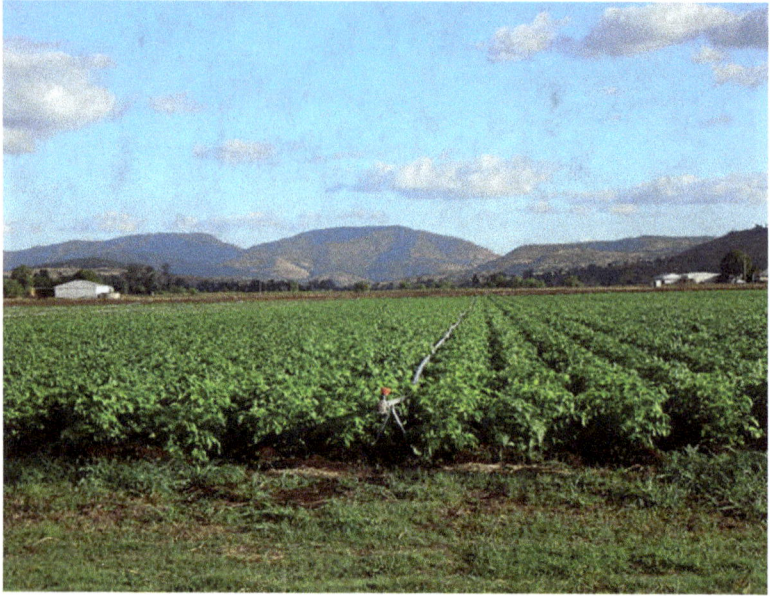

Figure 1.3   Efficient irrigation, Lockyer Valley, Queensland, Australia
(Photo courtesy of C Baldwin 2004)

Figure 1.4   Rainwater harvesting from school roof for greenhouse, Tana River catchment, Kenya
(Photo courtesy of C Baldwin 2012)

Figure 1.5   Mangroves stabilise Bells Creek bank, Australia
(Photo courtesy of C Baldwin 2011)

Figure 1.6   Hippos in Mara River waterhole
(Photo courtesy of C Baldwin 2012)

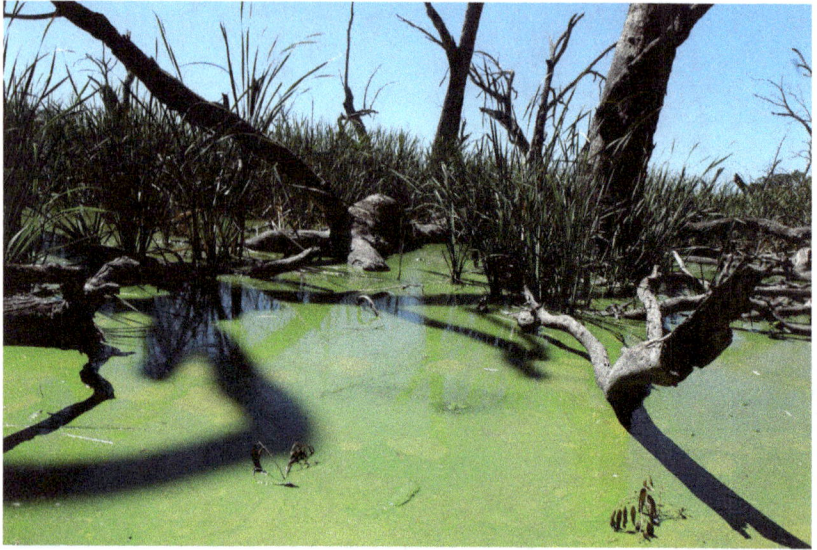

Figure 1.7 Blue-green algae bloom, Murray-Darling Basin, Australia

(Photo © Murray-Darling Basin Authority. Photographer Arthur Mostead. Used with permission.)

Figure 1.8 Eden Mills Millpond – sluggish summer flow in Grand River tributary from mill weir

(Photo courtesy of C Baldwin 2012)

Figure 1.9 Dry watercourse in Lower Balonne catchment, Australia, during drought
(Photo courtesy of C Baldwin 2005)

Figure 1.10 March St, town of Peebles, Scotland, UK during flood from Eddleston Water
2012
(Photo courtesy of C Spray, University of Dundee, July 2012)

Figure 1.11 March St, town of Peebles, Scotland, UK after flood from Eddleston Water 2012 (Photo courtesy of C Baldwin September 2012)

Figure 1.12 Field day illustrating sustainable rice intensification, near Lake Victoria, Kenya (Photo courtesy of C Baldwin 2012)

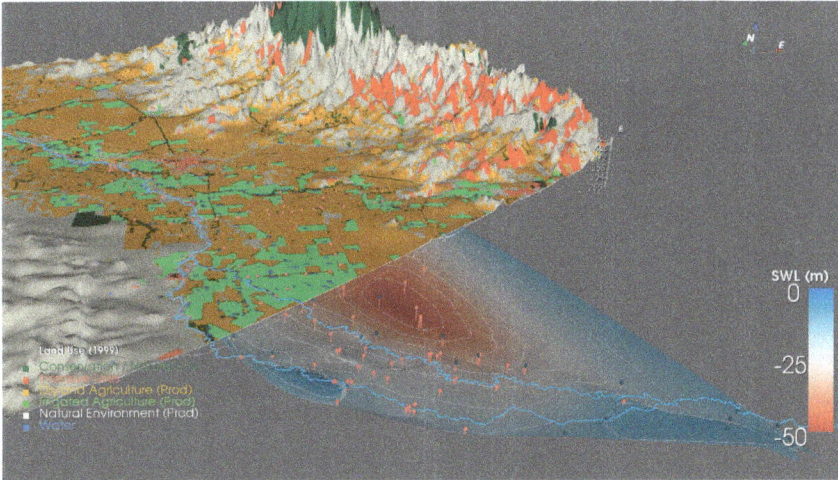

Figure 4.2 3D visualization of Condamine Valley in southeast Queensland, Australia, using GVS software, showing land use and alluvial groundwater levels in metres

(Source: unpublished report, A. James and M. Cox, Queensland University of Technology, 2013. Used with permission.)

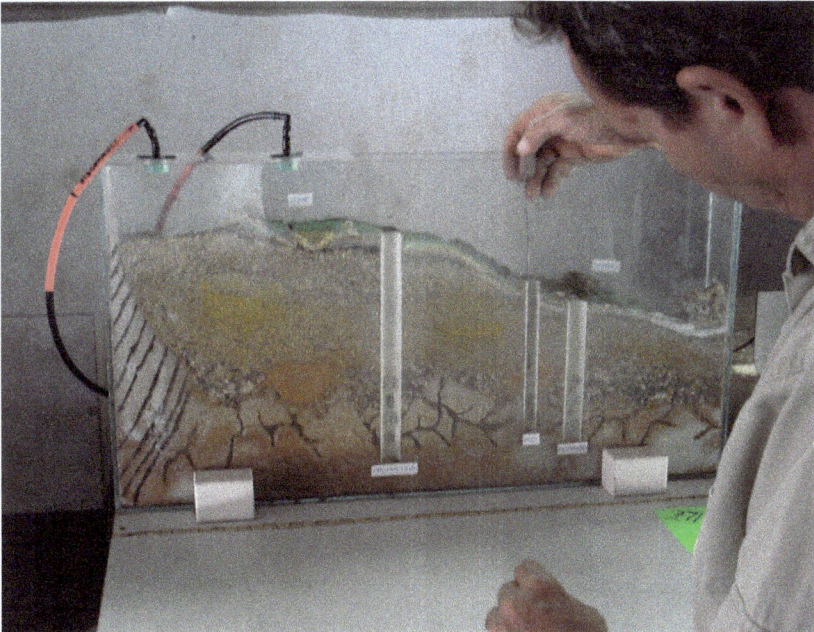

Figure 4.3 3D physical groundwater model

(Source: Northern Territory Natural Resources, Environment, The Arts and Sport. Used with permission.)

Figure 4.4 Community consultation issues flagged on map, Managing Borderlands project, UK

(Photo courtesy of J Forrester 2012)

Figure 4.5 Women in an impoverished community in Indonesia learning how to treat their drinking water

(Photo courtesy of Vikki Uhlmann Consultancies, Yayasan Emmanuel Water Program and Engineers without Borders (Australia))

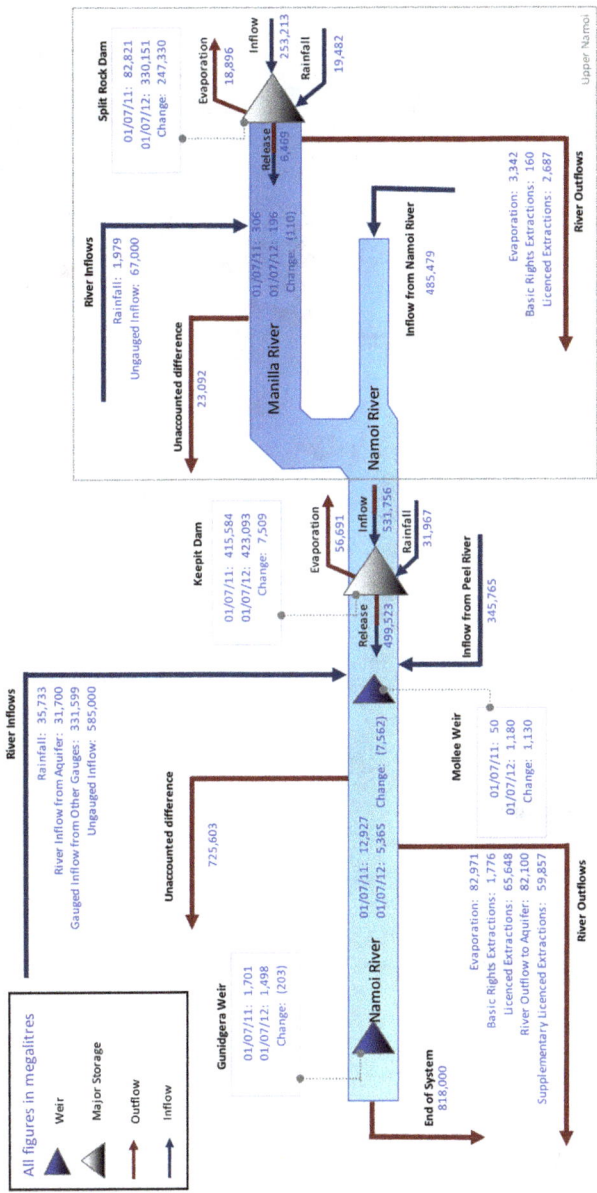

Figure 5.5 River system static water balance conceptual model for the Namoi River in Australia, with actual quantities for 2012–13

(Source: Burrell *et al*. 2013: 18 (NSW Office of Water General Purpose Water Accounting Report). Used with permission.)

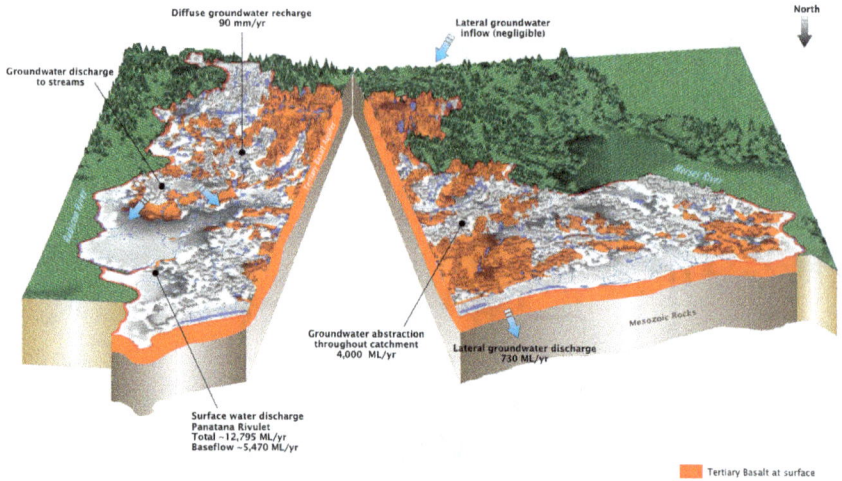

Figure 5.6 Groundwater resource static water balance conceptual model Wesley Vale, Tasmania, Australia, estimated long term average quantities

(Source: Harrington and Currie 2008: Figure 16,17. Used with permission.)

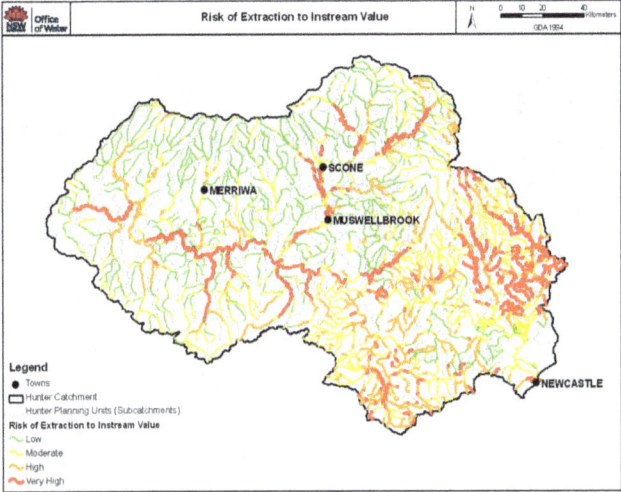

Figure 5.12 Risk of extraction to instream value: Hunter River Basin, NSW, Australia (Source: Hamstead 2010)

Figure 7.5 Trial to slow water flow in small creeks, River Tweed catchment, Scotland, UK (Photo courtesy of C Baldwin 2012)

Figure 7.6   Monitoring groundwater levels by University of Dundee to determine impact of meander interventions, River Tweed catchment, Scotland, UK

(Photo courtesy of C Baldwin 2012)

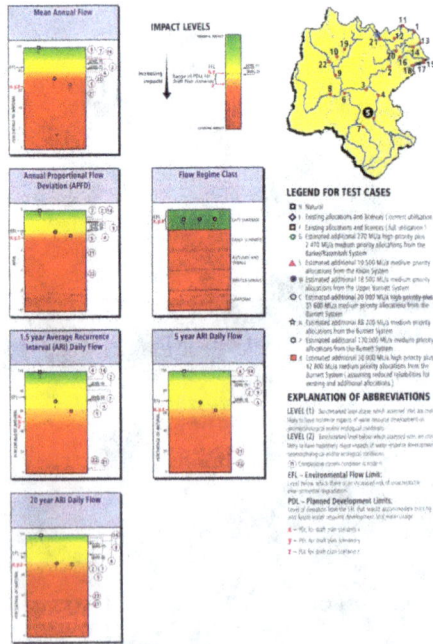

Figure 8.1   Burnett 'traffic lights' diagrams

(Source: QG 2000. Used with permission.)

	Code	Confidence flag	S/ML	kgCO2e/ML/yr	River Health	Acceptability	Volume (ML/yr)	Strategy S/ML
Operations/River Management								
Channel Automation Macalister Irrigation District - savings for environmental flows	E2						30,000	<50
Alteration of the Fyansford Quarry dewatering discharge licence from Corio Bay to the lower Moorabool River during times of low flow	E3		TBC	TBC			3,000	TBC
(Re)-allocation								
Provision of water to the environment – Yan Yean Reservoir	E7						3,000	50–150
Provision of water to the environment – Tarago Reservoir	E8						3,000	50–150
Assign the unallocated water in Blue Rock Reservoir to the environment	E9						31,000	50–150
Unallocated share of inflows into Lake Merrimu – for environmental flows	E10						1,800	<150
Provision of water to the environment – Rosslynne Reservoir	E12						3,000	<150
Reallocating irrigation water to the environment – 100ML capacity share in Rosslynne Reservoir	E13						100	<150
Provision of water to the environment – West Barwon Reservoir	E14						1,000	<150
Provision of water to the environment – Upper Yarra Reservoir	E15						17,000	50–150
Provision of water to the environment – Lal Lal Reservoir	E16						3,000	<150

Figure 8.2 Sample of summary table from Central SWS (Victoria, Australia) 'sustainability' assessment

(Source: Victorian Government 2006: 111. Used with permission.)

Option name:	Provision of water to the environment – 100ML capacity share in Rosslynne Reservoir	Option Code	E13
Description:	This is the difference between the annual entitlements of the irrigators and their capacity share in Rosslynne Reservoir. Would adversely impact on irrigators' reliability of supply unless the entitlement was traded.	$/ML	<150
		Ml/year	100
		Water Bill change	N/A
	This option is only at the conceptual stage. Cost range has assumed this option would pay a market rate for the allocation of entitlement. Also assumes no additional infrastructure requirements at the outlet of Rosslynne Reservoir. Would redistribute water from irrigators to the environment for no discernible river health benefits. Would face significant opposition from the community.		

Cost	Regional Development	Greenhouse Emissions	River Health	Water Quality	Other Ecosysems	Social Acceptability	Recreation & Heritage	Public Health	Confidence Flag	Fairness Flag

Figure 8.3 Sample of detail report from Central SWS (Victoria, Australia) 'sustainability' assessment

(Source: Victorian Government 2006: 120. Used with permission.)

Figure 9.1 Instream structure to improve accuracy of low flow monitoring, urban stream in Sydney, Australia.

(Photo courtesy of Mark Hamstead January 2014)

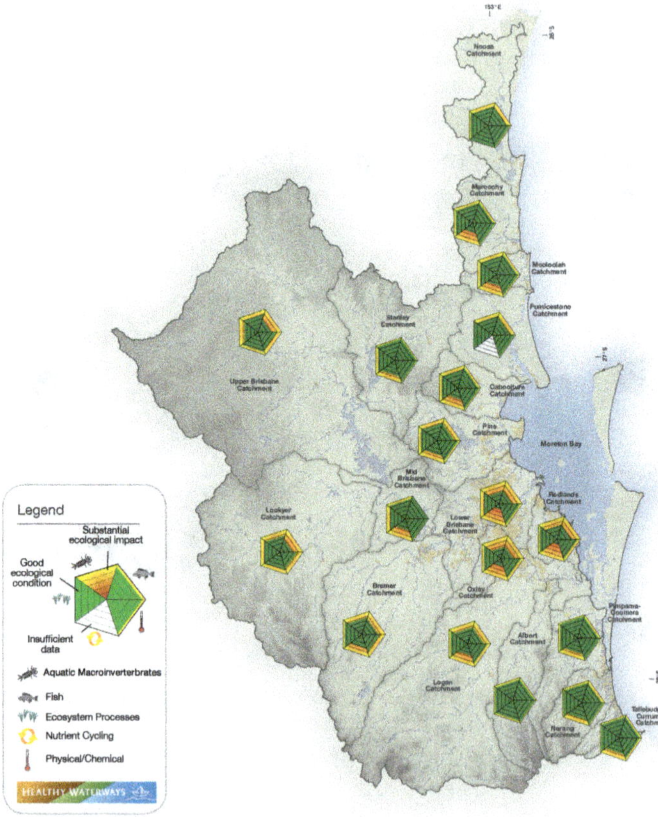

Figure 9.2 South East Queensland, Australia, Ecosystem Health Monitoring Program
results 2006–07

(Source: SEQHWP 2010:5; © Healthy Waterways Limited. Used with permission.)

7 Management options

In the *situational analysis* step of a water resource planning process, the benefits and beneficiaries related to the water resource were identified and the relationship between these benefits and the water resource characteristics considered. In the *objectives and logic* step desired benefits to be achieved were identified in objectives, and the associated outputs and performance indicators developed based on the knowledge from the situational analysis. Given this, the next step requires the identification and comparative assessment of management options and the determination of the actions to be used.

In this chapter we outline a range of different kinds of actions that have been used in different circumstances, and illustrate these actions with case studies. Sets of actions are commonly grouped together in *management options*. Management options typically vary in the extent that they achieve competing objectives and in implementation cost. In chapter 8 we discuss how management options are selected, compared, assessed and revised until a decision is made on which actions are to be used.

Possible actions

The kinds of actions that are in use are described below. Case studies are used to illustrate practical application in the remainder of this chapter. Whether they are addressed within a water resource plan, in broader legislation, or in associated planning processes varies from jurisdiction to jurisdiction. For example, legislation may dictate the nature and need for water licences generally, or empower plans to determine whether or not water licences are required in an area. The decision to build a dam and its parameters may be made by a government independent of water resource planning or may be the main subject of a water resource plan. Linkages between a government department responsible for water management and a catchment authority or water user association may be used to facilitate action outside of the normal legislative responsibilities.

Investing in new water supply infrastructure

The construction and operation of dams, weirs, irrigation channels, bores, pump installations, pipelines, etc., is a common way of improving reliability of access to water. This includes large dams that can be used to capture wet season flows for release in drier times to support irrigation, industry or towns; smaller capacity weirs on rivers that can provide water flow regulation as well as provide head to allow diversion of water; and channels and pipelines to carry water to towns and agricultural land. In *100 Ways to Manage Water for Smallholder Agriculture in Eastern and Southern Africa* are described some common and more innovative intervention techniques for capturing water especially relevant to developing countries (Mati 2007). Infrastructure can be proposed and funded by governments, aid agencies or private investment.

While the details of siting, design and construction of dams, weirs, etc., is usually dealt with by specific studies outside of water resource planning, a water resource plan may allocate a volume for this purpose, assess benefits and compare alternative locations and storage capacities taking account of environmental flow requirements and competing consumptive and non-consumptive human uses, and to develop multi-purpose operating rules.

Investing in new hydropower infrastructure

The construction and operation of hydropower dams and pipelines enables the generation of electricity. As with water supply infrastructure, water resource planning can be called upon to assess benefits and compare alternative locations and capacities taking account of environmental flow requirements and competing consumptive and non-consumptive human uses, and to develop multi-purpose operating rules.

Investing in water supply augmentation technologies

To meet or supplement consumptive water demand, and to ease pressure on existing water resources to achieve environmental or water security outcomes, cities, towns and some high value industries such as mines find it feasible to invest in alternative technologies for water supply. Examples are desalination plants, managed aquifer recharge, and water recycling (NWC 2013). A study comparing technologies is described in Box 8.4.

Regulation of existing infrastructure operation

A water resource plan may have within its scope the setting or revising of operating rules for dams and weirs, to better take account of environmental flow requirements and non-consumptive water uses that were not well considered in the past, as well as new consumptive water demands. The

Tennessee Valley Authority case study (see p. 169) illustrates its policy for operating the Tennessee River and reservoir system to meet multiple needs.

Investing in upgrading ageing infrastructure

Ageing urban and irrigation distribution systems in particular can be revitalised to reduce losses and improve operational efficiency. This can be particularly important in areas where demand for water exceeds supply as an alternative to building new dams, or to allow more water to be left for environmental and non-consumptive human uses without impacting on economic production. Case studies addressing ageing infrastructure are presented later in this chapter.

Direct regulation of water extraction

Individual water users can be required to comply with limits on the volume, rate and timing of extraction, usually through licensing arrangements. Licensing of water extraction requires that laws be in place that prohibit extraction of water except under an issued water licence, and that these laws are enforced. Such water licences can be issued to persons, companies, community groups and organisations, cities and towns, and even government agencies. The laws can provide for exemption from licensing or volumetric limitations in specific circumstances, e.g. for small-scale domestic drinking and sanitation or for firefighting. The integrity of the water licensing arrangement as a whole is dependent on there being no water extraction outside of the licensing system that could defeat its purpose, for example through large exemptions or poor enforcement.

Water licences can be defined in many different ways. They can be for a fixed term or perpetual. They can be restricted to extraction from one location or can be movable. Ownership can be fixed to the applicant or transferable. Limits and conditions on extraction can be defined by volume, purpose, rate, timing, authorised infrastructure or other parameters.

Water licensing is not always practical or cost effective to implement. As an alternative to water licensing, water extraction can be regulated through restrictions imposed generally or in an area, on a permanent or periodic basis, on such things as pump size, irrigated area and crop types, pump timing, etc. Such restrictions are proclaimed, publicised, explained and enforced under legislative powers.

Water resource plans may define the maximum volume of water able to be taken per season or year or during particular circumstances under licences as a whole, groups of licences of a kind, individual water licences or for water users in a particular area where there is no licensing system. These limits are used to preserve water for non-consumptive benefits (ecology condition, recreational water use, etc.), and for sharing of water between consumptive users. In the Padthaway groundwater area in Australia, limits were defined to manage salinity (see Padthaway case study below).

Day-to-day rules on the taking of water from rivers can also be designed to protect low flows and flow variability for the sake of downstream water users and non-consumptive benefits. They can include rules prohibiting or restricting pumping when river levels fall below trigger levels. Restrictions on volumes and extraction rates from groundwater in designated areas can be designed to control localised drawdown that could impact groundwater-dependent ecosystems and other water users. Rules can also be made to prohibit extraction from within a given distance of an ecological asset to protect water-dependent ecosystems.

Urban water restrictions

In most cities and towns water is supplied to individual homes and businesses by a water authority or 'water provider' which is responsible for managing within their licensed supply constraints. In such cases restrictions on purposes for which this water can be used are often applied during times of water shortage, including such things as prohibiting use of sprinklers to water gardens, alternating days for hand watering of gardens depending on street address, and limiting washing of cars and concrete surfaces.

Economic instruments

Economic instruments can be used to encourage desired behaviour. Examples are establishment of formal and informal markets in privately held rights to take water, subsidies/incentives for such things as on-farm water use efficient infrastructure (e.g. laser levelling, on-farm storage, improved irrigation reticulation equipment, water probes), rainwater tanks, water efficient appliances in homes, fees for water use or water pollution and grants and loans to assist in adjusting to change.

Markets in rights to water are a good means of allowing use of water to move around between uses where the water resource is fully committed. The fundamental underpinning is that rights to take water are encapsulated in licences or entitlements that can be moved in location and changed in ownership on either a temporary or permanent basis. They are not necessarily appropriate everywhere. There is a base cost to set up the necessary legislation, tradeable water rights, systems and procedures to administer trading, and optional costs to facilitate trading through such things as collating market information, operating exchanges, etc., that should be considered in deciding whether markets are a viable option.

A case study of water trading in the Murray-Darling, Australia is described below. Trading in water rights is usually highly restricted and localised, such as in the Orange–Senqu in South Africa, and in the Colorado River in the USA. On the Colorado, trade is not permitted between states, and 'within states the seniority system confers different values on each water right that complicate and limit trade' (Grafton 2013: 320).

Water markets have enabled use of a mechanism to achieve ecosystem outcomes – buying-back water for the environment, such as in the over-allocated Murray-Darling Basin (MDB) in Australia.

> In 2007, the national government earmarked A\$3.1 billion to buy water entitlements outright for the environment. By the end of May 2012, over half of this amount had been spent. About 1253 GL of water entitlements with a long-term average annual yield of 1031 GL (or about 8% of the total MDB entitlement stock) had been purchased, largely through competitive tenders (DSEWPC 2012, cited in Horne 2013: 533).

In some cases, those affected by a policy shift, press for 'structural adjustment'. For the most part adjustment assistance is discouraged as it can mask or distort long-term signals. If not well designed it can be inequitable and impede innovation. On the other hand it might be appropriate if an affected group is easily identifiable, has limited capacity to handle change, and where the impacts are well specified and clearly associated with the proposed policy change. The more successful of these approaches include an industry package that includes grants and loans for training to improve management skills or gain professional advice; and regional development grants for hard and soft infrastructure to improve the economic climate (McColl and Young 2005).

Improving institutional governance

Often the historic division of agency roles and responsibilities can impede effective water resource management. Arrangements to improve governance can include such things as improved coordination among state government agencies, multi-level relationships between local, state and national governments, and with catchment/river basin authorities; separation of conflicting roles such as water user and regulator; and more local delegation, for example by delegating powers and resources for water resource planning and management to a local or regional government level or to co-management bodies such as water users groups. Aspects of good governance are touched on in chapter two.

Facilitating local self-governance

Rather than externally imposed regulation or economic instruments, local communities can be supported to establish flexible self-governance arrangements, for example to share water in times of limited availability, or to improve water quality or ecological outcomes.

Box 7.1: Self-governance of irrigation systems in Australia

An investigation of self-management in relation to irrigation in Australia found several entities that were developed for the purpose of supplying surface water within irrigation schemes to members (Baldwin *et al.* 2008). They ranged from small (the 8-member Abercrombie Pumping Association in New South Wales) to large (the 1,500-member Central Irrigation Trust in South Australia). The self-managed entities distribute water as a bulk entitlement holder to water users via an infrastructure scheme (pipes, channels, etc.) according to defined entitlements and manage water trading within the scheme. Most of the entities had the ability to impose penalties and/or cut off supply if members did not comply with allocation rules. All of the case study entities were supported by fees or levies (some quite small) on irrigators, based on a flat membership fee, unit of entitlement, area used for irrigation, or volume of water used.

No cases were found where control of water sharing was delegated to a group of self-supply water users, i.e. those accessing water from groundwater or directly from rivers. However, in 2008 a group of groundwater irrigators from the Lockyer catchment in Southeast Queensland Australia proposed a system whereby sub-catchment groups of water users would be allocated a volumetric amount of water to be 'co-managed' with government within sustainable limits. Sub-catchment groups would develop their rules to apply within their group to enable flexible use and transfer of water within the system. The proposal included installation of meters by irrigators consistent with government specifications, regular monitoring by irrigators for feedback on use, and annual auditing by government (Sarker *et al.* 2009). The proposal was ultimately rejected by government on the basis of legislative and policy constraints. In addition to being self-empowering and gaining support of irrigators for new restrictions, it was an opportunity to build social capacity and meet the principles espoused throughout this book.

Models of self-governance draw heavily on Ostrom and colleagues' work and eight principles that lead to successful self-organised processes and cooperative behaviour (Ostrom *et al.* 1999; Ostrom 2005; Ostrom and Nagendra 2007). Essentially effective water resource management is more likely to occur if rules are based on sound data and developed in collaboration with users. Involving entities and members in monitoring extraction and resource condition provides continuous feedback about the resource. It is essential that users understand the system characteristics, impact of extraction, and uncertainties, so that an adaptive management approach is adopted. Government can legitimise and support a group's roles by recognising it in legislation, delegating powers, engaging it in development of plans, sharing data, and by providing support for training and transition activities. Such entities can also foster credibility and leverage resources by collaborating with government, research agencies, and regional natural resource management bodies.

Benefits of self-management entities have been well documented (Ostrom 1992, 2005; Marsden Jacob Associates 2004). They provide a mechanism to engage irrigators in developing rules for water sharing, recognising their expertise and knowledge, and gaining their commitment to management of a common water resource in an empowering way. Flexible management arrangements can be developed within the overall constraints of a water resource plan. If effective, self-management reduces the operational demands on government, facilitates commitment and innovation at a local level, and engenders a culture of compliance with rules that are seen to be fair and reasonable. Ostrom and colleagues' studies have shown that successful self-governing entities can have major roles in monitoring and compliance. Further, irrigators may also want control through metering, monitoring and compliance.

On the other hand, there are two main risks with self-governance. First, irrigators may not perceive that the transactional and financial costs and additional responsibilities of operating such an entity are worth the benefit, especially if they were engaged well by government in developing the 'rules' (through a water resource plan) in the first place. The second risk is that the entity does not meet its obligations under a plan or agreement and therefore does not meet government's legal or policy obligations. To address the risks and be successful, self-management needs to be formalised, rules (a plan) established in a transparent and inclusive way, and responsibilities accepted and funded within the capacity of the irrigators to support them (Baldwin *et al.*2008).

In the case of Lombok in Indonesia (see case study later in this chapter), it was found that though local rules were not always recorded, water user associations (WUAs) were formalised structures in the hierarchy of statutory mechanisms. This reinforcement of WUA legitimacy in legislation is also seen in other parts of the world. Kenya's Water Bill 2012 (s16) for example, specifies that:

> Water Resource Users Association[s] shall be community based associations for collaborative management of water resources and resolution of conflicts concerning the use of water resources. The Water Resource Users Association shall be established as [an] association of water resource users at the sub-basin level based on rules issued by the Water Resource Regulatory Authority.

In the case of the Andhra Pradesh state of India, the *Andhra Pradesh Management of Irrigation Systems Act 1997* is reported to be 'one of the most revolutionary legislations made by any government in the country' (ADB 2006: 51). The Act provides a framework for WUAs to be constituted with well-defined jurisdictions, role, and functions of farmers' organisations and the irrigation agency. It facilitates allocation of state funds to appropriately organised WUAs.

As illustrated, the legislative legitimacy for WUAs can encourage collaboration among irrigators, and between irrigators and government. However, to retain a sense of empowerment to manage the resource for the collective good, in keeping with Ostrom's principles, WUAs also need to have major input on decisions about rules that apply within their group, participate in monitoring, and compliance, even if at an informal dispute resolution level. Incentives for collaborative water management might include: more clearly defined entitlements; better understanding of reliability of supply (particularly in over-used systems); and tradeable water entitlements (Baldwin *et al.* 2008).

Investing in community education and capacity building

Closely related to self-governance, communities can be trained and provided with information and tools that enable them to better manage and use water and take care of ecosystems. Investing in educating water users in such things as how the water resource behaves; how use of water affects such things as water quality, fish and vegetation; the likelihood and extent of droughts and floods; interpreting monitoring information; and how to use water more efficiently can contribute to achieving objectives as they adjust their behaviour based on better understanding. Joint fact-finding was mentioned in Chapter 4 as a mechanism for community engagement as well as capacity building.

Investing in monitoring and research to improve knowledge and adaptive management

Where there is limited knowledge in matters important to water planning, investing in improving knowledge can be done through installing monitoring equipment, engaging communities to gather information, and engaging experts to gather information and conduct studies. For information that requires years to collect, an action in a water resource plan may be to collect this knowledge prior to a review of the plan in several years' time.

For example, where understanding of groundwater behaviour is limited, a government may install its own series of monitoring bores to measure fluctuation in water level, chemistry, and flow direction. Licence conditions may specify that individual licensees have a meter on their pump or bore to measure extraction. This may be used for compliance to volumetric licence conditions, however a major recognised benefit is for self-monitoring and self-regulation of water use to improve water use efficiency. Metering can also be used to determine where water losses occur. Similarly a plan may recommend the installation of river gauges to gather information needed to better understand the behaviour of the water resource over time. Experts can be commissioned to identify, characterise and map groundwater-dependent ecosystems and the services and benefits they provide; and gather information about on-farm irrigation practices.

Box 7.2: Benefit of water accounting in Sri Lanka

Water accounting tools are useful in characterising water use, pointing to where greater in-depth studies are required, and in planning and developing strategies for water savings and increasing water productivity.

From measuring rainfall, flow and other factors, it was found that after passing through a cascade system of inter-connected tanks (reservoirs), 92% of the water was depleted through evaporation and transpiration. Only 3% of the gross inflow into the catchment was actually delivered for crops. The analysis suggested ways to increase productivity of water by addressing storages, cropping intensity, and even maintaining forest cover at a watershed scale (Molden and Sakthivadivel 1999).

Trigger response mechanisms

Where there is risk but a lower likelihood of a threat occurring, one approach is to delay action, but monitor, so that should the threat occur, a response is triggered. For example, actions in a plan may be designed to maintain groundwater levels above a certain point to ensure that there is no saline intrusion from an adjoining saline aquifer. Uncertainty about how the aquifer will respond to extraction and future recharge can warrant inclusion in the plan of 'trigger-response' mechanisms, so that if the aquifer water level declines below a defined level, restrictions on pumping and a review of pumping rules is triggered. In Tasmania, the Ringarooma Water Allocation plan includes a requirement to monitor certain river health parameters and to review plan actions should they significantly decline (DPIPWE, 2012: 23).

Adaptively managed water allocations for the environment

Where there are dams and weirs regulating flow on rivers, and issued rights to water captured in those structures, traditionally those rights have been issued to consumptive users for irrigation, town water supply or industrial/commercial purposes. These rights allow the holder to call on capture water at times when they need it. A more recent innovation is for entities to hold and use such rights for environmental purposes.

In the larger river systems of the Murray-Darling Basin in Australia, an increasing volume of rights to water have been re-allocated from consumptive use to entities that use the water for environmental water purposes, to address declining river and wetland health. Most of this has been done by purchase of water rights from irrigators by government agencies over the last five years. These 'environmental water entitlements' are held by statutory government bodies such as the Commonwealth Environmental Water Holder and the Victorian Environmental Water Holder. Governance arrangements provide for adaptive use of the water to achieve environmental outcomes such as increasing flooding frequency and duration to highly valued wetlands that have deteriorated.

Case study – Northern Region Sustainable Water Strategy, Victoria, Australia

Northern Victoria sits at the southern end of the Murray-Darling Basin in Australia. The Sustainable Water Strategy (SWS) covers a group of rivers that flow north into the River Murray (Figure 7.1). The rivers and aquifers support a major agricultural region which generates (in Australian dollars) around $3.26 billion per year in production value, including $924 million from irrigated horticulture and $707 million from dairy (Victorian Government 2008b: 28). The water resources also supply drinking water for large towns, villages and farms, as well as providing water-based recreational activities.

Water infrastructure includes eight large reservoirs and several smaller reservoirs (total storage capacity approximately 11,500 gigalitres), and 6770 km of irrigation channels, most of which were constructed decades or longer ago. There are no worthwhile opportunities for further substantial reservoirs. Infrastructure maintenance and operation are funded largely through fees on water use.

Water rights are well established, volumetrically defined, metered and enforced, with a highly developed system of tradeable water entitlements and markets in those entitlements. Costs are recovered through fees on transactions and water use.

The region is also home to iconic water-dependent plants and animals, as well as many significant wetlands, three of which are Ramsar-listed.

Faced with a record drought and future climate projections suggesting long-term reductions in water inflows, from 2007 to 2009 the state government developed a water resource plan for the region – called a Sustainable Water Strategy – that sought to address future risks to the local economy and to valued ecosystem assets.

In a discussion paper in January 2008 (Victorian Government 2008a), the state government set out the situation, set out benefits currently arising from the water resources, projections of future climate and risks to those benefits under current management arrangements. The paper also outlined several kinds of actions that could be undertaken, and sought community input on

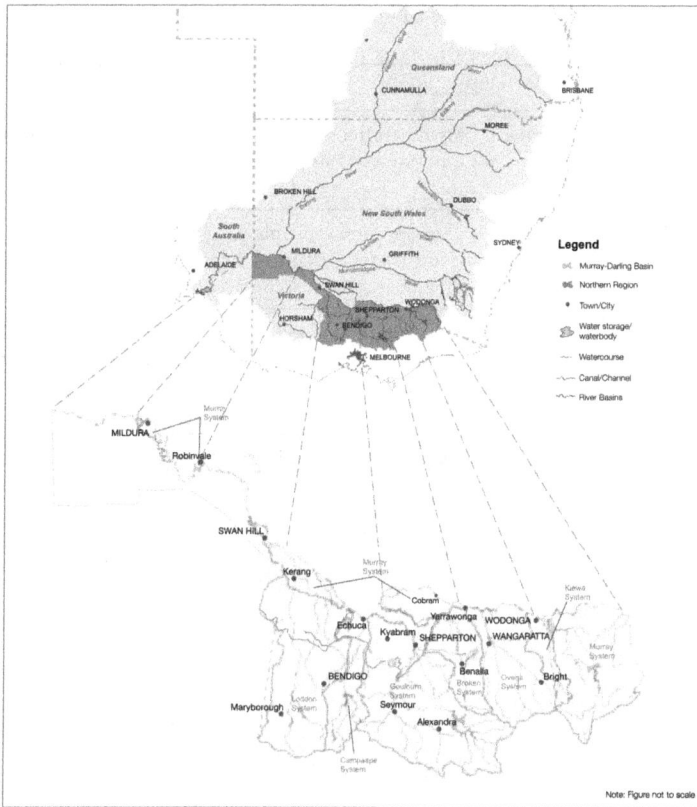

Figure 7.1   Northern Region Sustainable Water Strategy, Victoria, Australia
(Source: Victorian Government 2008b: 21. Used with permission.)

future directions, which actions were thought to be most important and other actions that could be done.

In its draft strategy published in October 2008 (Victorian Government 2008b), the government explained proposed actions aimed to address risks to environmental assets and essential water supplies, and provide irrigators with greater capacity to adjust to a projected future with much lower water availability. Many of the proposed actions are set out in Table 7.1. As shown, each option has both costs and benefits.

Case study – Burnett Basin Water Resource Plan, Queensland, Australia

The Burnett Basin Water Resource Plan covers approximately 38,370 km^2 in South East Queensland, Australia (Figure 7.2). Several dams and weirs within the basin provide water supplies that are largely used for irrigated agriculture, but also support towns, power generation and mining.

Table 7.1   Options considered in developing the Northern Victorian Sustainable Water Strategy

Action	Benefit	Cost
Increase the amount of reserve retained in storages at the end of each year	Provide greater security of supply to water users in dry times	Reduce water availability in average years
Shorten the irrigation season in very dry years, to reduce losses of water in channel systems	Increased water available for towns and agriculture in subsequent years	Reduced agricultural production at the end of dry years
Allow water entitlement holders to carry forward to the next year a portion of unused annual water allocations	Greater ability for water users to self-manage risk and reduce end of year low value usage	Reduced water available for allocation to all next year
Add pipelines to connect towns with additional water supply sources	Towns have more options for water during droughts	Financial cost of infrastructure
Allow water entitlements to be dedicated to environmental purposes, and provide efficient governance arrangements	Improved condition of water-dependent ecosystems	Cost of governance
Modernise ageing irrigation channel systems to reduce losses of water and use savings to create environmental water entitlements	Improved condition of water-dependent ecosystems	Financial cost of infrastructure
Purchase irrigation water entitlements and dedicate them to environmental purposes	Improved condition of water-dependent ecosystems	Reduced agricultural production; financial cost of purchasing water entitlements
Adjust rules for operation of dams and weirs to achieve multiple benefits above and beyond releases for irrigation supply	Improved condition of water-dependent ecosystems, improved water quality, greater security for tourism and recreation benefits	Reduced water for irrigation
Reduce constraints on trading water entitlements	Individual water users have greater flexibility to manage risk; Overall economic benefits increased	Greater risk of third party and environmental impacts due to movement of water entitlements
Revise town drought response plans to incorporate additional worst-case scenarios	More certainty of supply for towns	Increased cost to have in place emergency measures for worst-case drought scenarios
Construct works to artificially increase flows into wetlands	Improved condition of wetlands	Financial cost of infrastructure construction, maintenance and operation

The Burnett Basin features a steadily growing population distributed in several townships. The population includes a significant Indigenous component, with areas of cultural significance located throughout the basin.

Agricultural production is the main economic activity in the basin with sugar cane being the most significant agricultural crop and water user in the region. Horticultural and grazing industries are also significant, with indications suggesting that growth in the tree cropping sector (for example, citrus and macadamias) is likely to occur in the future.

Other industries in the basin include some services and industrial activity, such as mining, Tarong power station and food processing plants based in

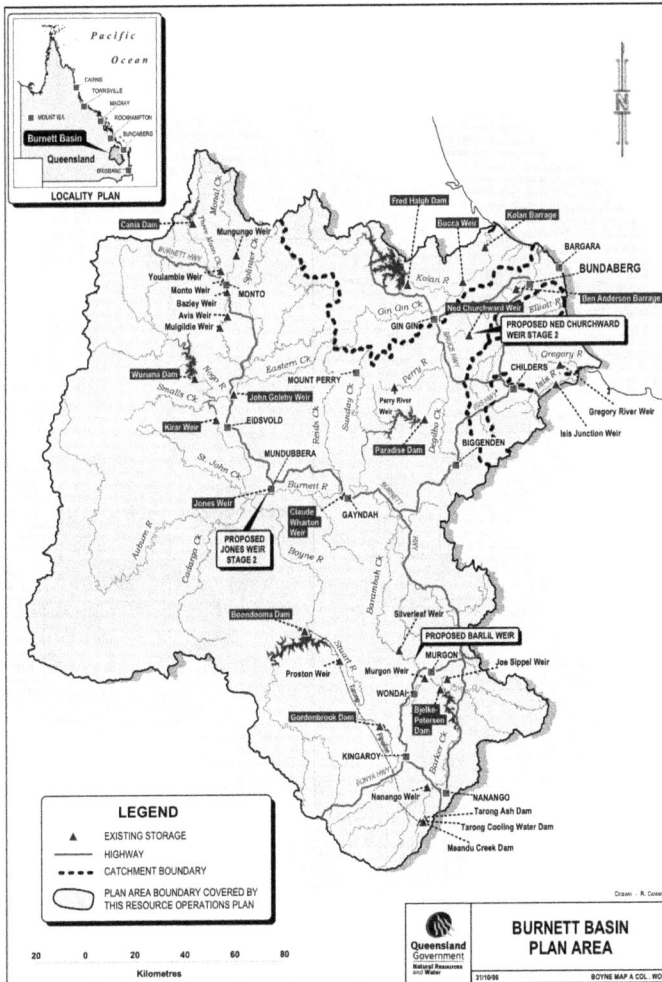

Figure 7.2 Burnett plan area
(Source: QG 2006: 57. Used with permission.)

major population centres. Non-consumptive uses of the water resources in the Burnett Basin support economic industries, such as tourism and commercial fishing, and recreational activities for local residents.

From 1998 to 2000 the Queensland state government developed a water resource plan for the major rivers in the basin. The Water Resource Plan defines the overall limits on extraction and set objectives and management targets for environmental flows and water supply security. It was followed by development of a Resource Operations Plan (ROP) that defined in some detail the management actions that would be applied to achieve the defined objectives and targets (Table 7.2).

Table 7.2   Options considered in developing the Burnett Water Resource Plan

Action	Benefit	Cost
Government build new dams and subsequently issue new water entitlements	Greater economic production, recreational benefits of reservoirs	Financial cost; Risk to ecosystems
Requirement to operate dams and weirs to achieve low- and medium-flow targets, limit rates of rise and fall downstream, and comply with minimum water levels for recreational purposes and aquatic refuge	Reduced risk of loss of recreational benefits and damage to aquatic ecosystems	Lower immediate economic benefits
Build and operate fishways on dams and weirs	Reduced risk to fish populations	Financial cost
Barrage operation rules to prevent saltwater intrusion	Reduced loss of economic production and ecosystems due to salinisation of lower river	Financial cost
Conversion of existing water extraction authorisations to tradeable water allocations	Individual water users have greater flexibility to manage for risk Overall economic benefits increased	Financial cost of administration; Greater risk of third party and environmental impacts due to movement of water entitlements
Enforceable annual allocation volumes for water entitlements according to specified rules and formulae designed to allow for future reserves and water for the environment	Improved water security for towns and economic production in droughts; Reduced risk to ecosystems	Financial cost of administration and enforcement
Limit extraction of water from low flows in rivers without dams or weirs through statutory conditions	Reduced risk to ecosystems	Financial cost of administration and enforcement

Case study – Padthaway Prescribed Wells Area, South Australia

The Padthaway Prescribed Wells Area (PWA) is located approximately 300 km southeast of Adelaide and covers an area of approximately 700 km². The climate in the Padthaway PWA is hot, dry summers and cool, wet winters with average annual rainfall of around 500mm.

Agriculture is the main economic activity in the area, with a high reliance on groundwater for irrigation. The principal irrigated industry is viticulture, both in terms of area (3,110 ha) and, more significantly, economic value. There are substantial areas of irrigated pasture (2,110 ha), lucerne for seed production (732 ha), cereals (570 ha), coriander seed (444 ha) and canola (308 ha). Most of the remaining areas are made up of lucerne for hay or grazing and pasture seed (clovers). Other groundwater uses include town and rural domestic water supplies, but total usage is small compared to that used for irrigation (South Australian Government 2001: 1).

Water use in the area has been subject to licence for many years. Irrigation licences were originally issued to individuals for specified crop types and land areas, rather than being defined volumetrically. While the government had stopped issuing new licences prior to 2000, substantial potential remained for development under licences already issued. Transfer of water licences to different properties had also been enabled prior to 2000.

In the main irrigated area, the salinity of the groundwater has been rising for over twenty years. Further, assessments in 2000 indicated that full development of all issued licences would result in further increases in salinity and drawdown of aquifers. Consequently, a statutory water resource plan was prepared in 2000 that essentially put in place measures to prevent growth in risk and to improve knowledge (South Australian Government 2001) (see Table 7.3).

Over a seven year period from 2002, the water resource was managed in accordance with the limitations set out in the 2001 plan, and the actions to improve knowledge were implemented. Review of the plan commenced in

Table 7.3 Actions in the Padthaway Prescribed Wells Area Water Allocation Plan 2000

Action	Benefit	Cost
Restrict further development of irrigation through legally enforceable restriction notices issued to licence holders	Reduced risk of increasing salinity rises, and potential for irrigators to invest in infrastructure that might need to be retired	Loss of growth in economic production
Convert irrigation water licences from an authorised area to an authorised volume and install meters	Improve knowledge of water extraction and better defined licence holder entitlements	Financial cost of metering and conversion process
Undertake more rigorous assessments of the capacity of the aquifer to support water extraction	Improved basis for decision-making	Financial cost of assessments

Table 7.4   Actions in the Padthaway Prescribed Wells Area Water Allocation Plan 2009

Action	Benefit	Cost
Permanently reduce allocations in licences through action under statutory powers	Prevent future water users from being impacted by rising groundwater salinity and declining water tables	Loss of growth in economic production
Impose trading rules that prevent localised increases in water extraction in critical areas	(As above)	Individuals in critical areas prevented from expanding development
Differentiate on licences used for flood irrigation between water that returns to the aquifer and water that is evapotranspired, so that the returned portion cannot be transferred to a use where it is not returned	(As above)	Differentiation is an estimation and may not reflect the actuality
Allow water users to carry over unused water allocation from one year to the next	Greater ability for water users to self-manage risk and reduce end-of-year low value usage	Financial cost of more sophisticated allocation tracking and enforcement systems
Impose setback requirements for new wells from wetlands and existing wells	Protect wetlands and existing water users from being impacted by others	Possible that some new developments need to pipe water over larger distances
Providing for artificial aquifer recharge and subsequent water recovery (dependent on investment)	Potential for additional improvements in groundwater salinity, and ability for entities to use the aquifer as water storage	Financial cost for those who choose to invest in infrastructure, and of systems for tracking and enforcing limits

2005, with studies showing a need to reduce the levels of water allocation or risk continual increases in salinity. The planning process included extensive consultation with the local community and further studies, culminating in a revised plan in 2009 (South Australian Government 2009) (Table 7.4).

Case study – Rehabilitating aging infrastructure in Canada and India

Urban infrastructure – Canada

Canada, a relatively young country by global standards, is already starting to invest in repair and replacement of deteriorating infrastructure. According to a report by the Federation of Canadian Municipalities in 2007, water infrastructure in larger, older cities is particularly in need of rehabilitation. In

Montreal, 33 per cent of the water distribution pipes reached the end of their service lives in 2002; another 34 per cent of the water-pipe stock will reach the same state by 2020. St. John's pipelines are about three hundred years old. The study showed that the deficit related to the water supply, wastewater and stormwater systems across Canada in 2007 stood at $31 billion for the existing capital water infrastructure stock alone (Mirza 2007).

Replacement on this scale requires funds greater than those of individual municipalities. While federal funding has been allocated regularly to infrastructure works, a kick-start was facilitated through the $4 billion Infrastructure Stimulus Fund in 2009, which provided 50% federal level funding for projects matched by provincial and municipal authorities (Doyle 2009).

Deterioration and renewal of irrigation tanks in India

India has an estimated 350,000 small scale artificial tanks (reservoirs) and ponds, that have been used for centuries to store monsoon rainwater for multiple purposes: small-scale irrigation, for drinking water, bathing and sacred rituals, and aquaculture, depending on the region (ADB 2006). In the semi-arid regions of India alone, around 120,000 small scale tanks irrigate about 4.12 million ha (Anbumozhi *et al.* 2001). Use of tanks for irrigation declined rapidly after Independence, when canal irrigation was extensively developed. It took over some tank-irrigated areas and was accompanied by a breakdown of traditional local institutions that managed tanks. Over time, tanks and areas irrigated by tanks declined due to silting of feeder channels, reduction in storage capacity from siltation and encroachments in the tank bed, inadequate maintenance resulting in deteriorating tank structures, and heavy seepage losses in the delivery system. This has resulted in a reduction in supply to farms. Along with budgetary constraints for maintaining large irrigation schemes and few new major irrigation opportunities available, it became clear that rehabilitating the many tanks would be an important way for food production to match the needs of a growing population.

Over the past three decades a tank regeneration movement has resulted in rehabilitation of tanks primarily in India's drier south, funded by overseas aid, State governments, non-governmental organisations, and local communities. The ADB (2006: 64) reports that the investments in tank rehabilitation projects could be grouped under the following six items:

1 works to augment water to the tank
2 works to make the tank store more water and for a longer duration
3 works to ensure proper distribution of tank
4 water and minimise transmission losses
5 works to strengthen the tank users' institution
6 administrative cost for rehabilitation.

While the number of tanks rehabilitated effectively is negligible compared to the total number of tanks, primarily landholders and landless agricultural labourers (wage-earners) have benefited. Farmers who owned land in the area had the right to use the water and built, owned, and managed tanks, as a common property resource, through community organisations. A range of institutions and tank rehabilitation programmes were compared. It was found that when both livelihood and agricultural production were objectives for rehabilitation, the mean per capita income of villages increased by 66% (ADB 2006). Furthermore it appears to be a cost-effective option compared to creating new irrigation works. Consideration is also being given to integrating tanks into a larger waterscape linking them with larger dams so they are more resilient. This way the local systems act as rain-fed tanks in their own right but also as distribution mechanisms for the larger systems (Paranjape *et al.* 2010)

Case study – Regulations, incentives, restrictions and demand management in Southeast Queensland, Australia

As a result of drought conditions and low water reservoir levels in Southeast Queensland (SEQ), Australia, by 2008 a number of measures were put in place to manage water demand in urban centres. Reducing demand on reticulated supply requires some combination of either reducing overall consumption (demand management) and/or using alternative local supplies (such as re-use and rainwater tanks).

Rainwater tanks

In 2008 a Queensland Development Code MP4.2 was amended to require all new detached residential dwellings to have rainwater tanks of a minimum 5000 litres to reduce demand on the reticulated supply. In addition, subsidies offered by the State government to encourage installation of tanks in existing and new buildings were estimated to have contributed to an uptake of tanks of between 25–40% at detached and semi-detached dwellings in SEQ (QWC 2009a; White 2009). While 78% of SEQ homes were considered suitable for such installation (Marsden Jacobs 2007b: 22), cost to the homeowner was found to be the greatest inhibitor (White 2009). Highest efficiencies are achieved when plumbed to internal services (WBM Oceanics 2006: Marsden Jacobs 2007a), and so a second tranche of the incentive required tanks to be connected to non-potable services (toilets, laundry). When introducing requirements for rainwater tanks, specifications need to be location-specific due to the considerable variability in rainfall and therefore yield across a region (Warrick 2009). Supply from rainwater tanks is less vulnerable to climate change than streamflow or runoff (Coombes and Barry 2008). Acknowledging the need for appropriate management for mosquitoes and water quality and requirements for roof and other collection surfaces, rainwater tanks may go some way to solving issues of water supply in rural areas and developing countries.

Demand management

Existing literature suggests that demand management should be the first response to drought, climate change, and a growing population: it is the least costly solution, saves energy and reduces emissions (Marsden Jacobs 2007a; Turner *et al.* 2007). In the case of SEQ, restrictions were introduced and adopted quickly by residents with little controversy. Even two months after restrictions were lifted from 140 to 170 litres per person per day, residential consumption remained low at 129 litres per person per day in September 2008, with reported use a year later at 164 litres per person per day, well under the then restrictions of 200 litres per person per day (QWC 2009b). It was estimated that a lower level of supply could delay the need for costly supply augmentations, such as a desalination plant, by five years from 2017 to 2024. This lower use compares more favourably with, for example, average European water consumption of less than 150 litres per person per day (UNEP 2005).

In addition, water-intensive businesses were required to prepare Water Efficiency Management Plans to demonstrate they use water efficiently or to show how they plan to reduce their water consumption by a minimum of 25% in the future. Ninety-nine per cent of these businesses had submitted plans by 2009 (QG 2009).

Case study – Water user associations in Lombok, Indonesia

To meet Indonesia's Millennium Development Goals for potable water and livelihood improvement, allocation of water needs to be well managed to address the competing interests of agriculture, tourism and urban water in regional areas. In Lombok, good quality naturally flowing spring water from around the base of Mt Rinjani in Lombok is a significant resource providing drinking water to local villages and towns, the city of Mataram, increasing tourism development and bottled water companies. Springs are also the source of river water which is diverted for irrigation, mainly for rice, to sustain the population. However one-third of the natural springs have dried up (Transform *et al.* 2005). While long-standing processes are in place for distributing irrigation water through Water User Associations (WUAs), these additional competing demands threaten maintenance of sustainable yields and environmental flow of water sources and watercourses. Moreover local populations near springs feel powerless to maintain stewardship over the resource.

WUAs are democratically formed associations of water-using farmers that are recognised by government as having a legitimate role in managing water for irrigation. Many WUAs successfully manage water for irrigation through a traditional way of community decision-making (musyawarah), which aims for consensus (Klock and Sjah 2011; Sjah 2007). Participants explain their needs and the rationale, and then attempt to reach agreements that they feel are fair for all, by giving away some rights while receiving some benefits. Agreed rules for sharing water are generally but not consistently recorded in local laws (awig-awig). Some WUAs though are concerned about drying springs and water being allocated to the regional government water distribution company and bottled water companies for potable water, leaving local users feeling vulnerable to an increasingly unreliable water source. While government endorses the community-based WUA system, water allocation is also subject to issues of national importance and other regulations and laws. Hence, legitimate other users are also given permission to access the good quality spring water.

Baldwin and Sjah (2012) argue that applying Ostrom's principles tailored to the local situation could improve outcomes. One suggestion is to engage WUAs in 'payment for environmental (or ecosystem) services' (PES) schemes between upstream providers and downstream beneficiaries. Another is to apply the tradition and skills associated with musyawarah used by WUAs to multiple levels of decision-making that include WUAs as well as government agencies.

Case study – Water trading in Australia

The term water trading as used in Australia refers to the buying, selling and movement of rights to take water from a water resource, rather than the buying or selling of water itself. It can involve change in ownership and/or change in the location of extraction under the water right. For example:

- a person can buy land and at the same time buy the water right currently associated with that land. In this case an ownership transfer occurs but there is no change in the location of extraction;
- a person can move a water right they own from one location to another some distance away. In this case there is a change in location of extraction but no change in ownership;
- a person can buy a water right and move it to take water at a different location. In this case both an ownership transfer and change of location occurs.

Australia's National Water Commission (NWC) has commissioned several reports and studies about water trading in Australia. According to NWC (2011: 18), water markets stem from the basic idea of a 'cap and trade' system in which:

- the cap represents the total pool of the resource available for extraction, consistent with sustainable levels of extraction;

- individual users are provided with entitlements to a share of the total pool;
- entitlement rights and the quantity of water allocated to an entitlement each season (a water allocation) are tradeable, so that ownership, control and use can change over time;
- the price is determined in the market by the value placed on water by many buyers and sellers.

A fundamental aspect for establishing water markets is defining the water resource unit within which a 'cap' is set and trading can occur. The water resource unit is usually a river system or aquifer that is strongly internally hydraulically connected, so that movement of water extraction within the resource has a neutral effect on overall water availability. As a result, in Australia there is no single water market, but rather hundreds of markets defined by water resource units.

NWC (2011: 11) suggests that investment in facilitating markets will be most beneficial where:

- the water resource is fully developed with respect to consumptive use
- there is high variability in water availability on a seasonal and/or annual basis
- there are a large number of water users
- users have varying demands and degrees of flexibility to respond to water shortages
- water users are exposed to the cycles of global agricultural markets
- demands for urban and environmental water are increasing
- there is pressure for change in the existing structure of water-using industries.

While water trading is widely enabled in Australia, in most cases trading activity is low, because the units are small, there are only a small number of water rights holders, and/or there is little competition for the available water. NWC (2011: 25, 103) reports that over 90% of water trading activity in Australia occurs in the connected perennial rivers in the southern Murray-Darling Basin. These rivers consist of the River Murray and its major tributaries each of which has large dams to capture wet season flows. The dominant water use is irrigation, predominantly for dairy, rice and horticulture. All the criteria listed above apply to this area. The connected rivers do, however, span three states each with different systems for water entitlements.

The volume of water allocation and entitlement trading has grown significantly since the introduction of trading (Figure 7.3). The first significant increase was in 1994–95, when there was a large drop in seasonal water availability for the first time since the introduction of water trading. The second boost to water trading occurred in 2002–03. The severe drought in that year prompted a step-change increase in the proportion of water allocations that were traded, from around 7% to almost 15% (Figure 7.4). Water trading in this area reached an annual turnover of $ AUD 3 billion in 2009–10.

This expansion in trading has been facilitated by a series of government reforms over the last twenty years, including:

- moratoriums on issue of new water entitlements;
- recognising that full development of issued entitlements would exceed sustainable levels of extraction, setting caps on extraction at levels below full development, and requiring action to limit that development;
- firming up 'caps' on water extraction so that they included all kinds of water extraction in a catchment, not just licensed extraction from major rivers;

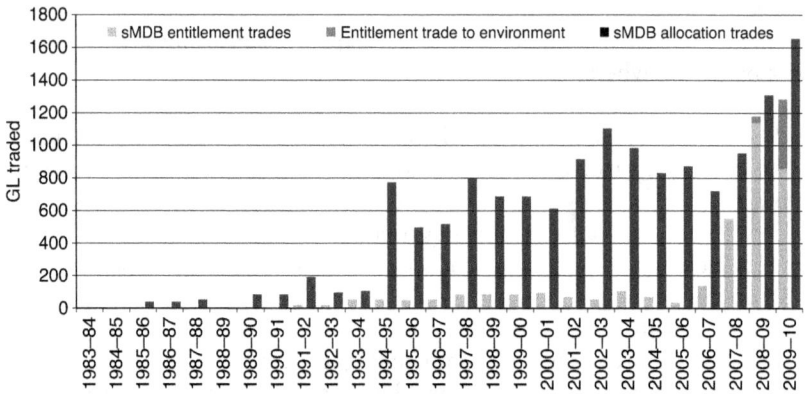

Figure 7.3   Volumes of allocation and entitlement trades in the southern Murray-Darling Basin, 1983–84 to 2009–10

(Source: NWC 2011b: 103. Used with permission.)

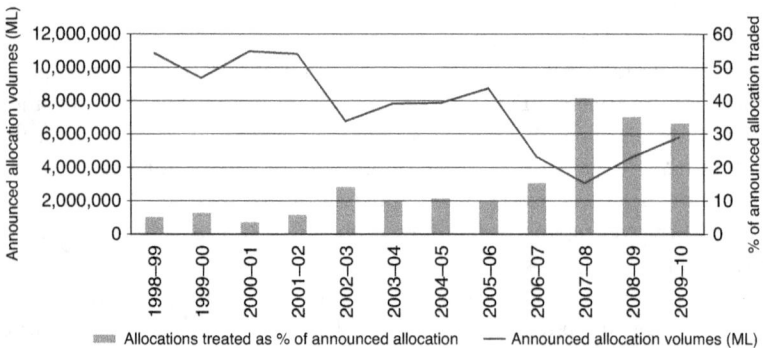

Figure 7.4   Water allocation sales as a percentage of water allocated in the southern Murray-Darling Basin, 1998–99 to 2009–10

(Source: NWC 2011b: 103. Used with permission.)

- development of water resource plans to implement the caps;
- changing water entitlements from an authority to install and operate pumps, bores, etc., and use water to being a perpetual share of available water in a specified water resource, able to be owned separately from land;
- facilitation of exchanges to help buyers and sellers come together;
- establishment of public registers of water entitlements and improving public access to information on prices paid;
- removing rules that impede trade for reasons that were considered to be not in the broader public interest (e.g. restricting ownership to water users);
- reducing costs and times for processing trading transactions;
- addressing administrative challenges associated with trading across interstate boundaries within the connected river system.

NWC (2010: iv–xii) reports a study that showed how water trading benefited irrigators in all major irrigation industries in the southern Murray-Darling Basin:

- *Rice growing:* Drought led to a major decrease in water availability for irrigation in the rice-growing areas. To generate income during the drought, rice irrigators typically sold their relatively small water allocations and further reduced or ceased annual rice production.
- *Dairying*: Water trading was a central element of many strategies adopted by dairy irrigators to deal with drought. For example, when water prices were high, allocation sales generated income that was used to purchase additional fodder. Alternatively, when prices were lower, some irrigators bought additional water to maintain production and capacity. Entitlement sales increased over the period as dairy farmers sought to manage debt and recent declines in commodity prices, and as they shifted from perennial to more opportunistic annual pastures.
- *Horticulture*: Wine grape and other horticultural irrigators in many areas were major purchasers of allocations and entitlements during the drought. Without water trading, it is likely that many long-lived horticultural assets would have been lost, at great cost to individuals and communities. As a result of declines in wine grape prices irrigators, particularly those on small blocks in public irrigation districts, are using water entitlement trading to help them exit the wine industry.

Many water users were concerned that water trading could result in mass movements of water entitlements (and consequently reduction of economic activity) out of local areas. The NWC study showed that while there was some evidence of this in a couple of areas, the broader regional economic benefits far outweighed them.

Groundwater trading in Australia is also occurring, but like surface water trading is scattered with the majority occurring in a few large aquifer systems. GHD *et al.* (2011: 27) report that of a total of 1284 trading transactions for a total volume of 308,019 ML in 2008–09, 55% by number and 75% by volume occurred in three large alluvial aquifer systems, where there are comparatively large numbers of water users, and demand for water exceeds availability.

In regard to further development of groundwater trading in Australia, GHD *et al.* (2011: 51) recommend that government investment in developing markets be commensurate with potential market activity. The kinds of investments they flag include:

- different levels of sophistication of water entitlements and supporting systems such as registers;
- assessments to improve understanding of sustainable extraction levels and physical system behaviour, as opposed to adopting conservative assumptions;
- compiling and publishing information on market activity;
- monitoring and compliance to reduce illegal extraction of water.

Overall NWC (2011: 21) report that water trading promotes economic efficiency by enabling water resources to be reallocated to those who value them most highly in both the long and the short term:

- Seasonal water trading enables the water available in any given season to be reallocated across crops, locations, irrigators and other water users in response to seasonal conditions (the concept of 'allocative efficiency'). This is particularly valuable where different users have different water demands. For example, given adequate forewarning, rice growers can simply choose to reduce the areas they sow during times of low water availability. However, other farmers, such as those growing perennial horticultural crops (such as fruit trees), need water every year. Trading provides the opportunity to move water between users with different water demands.
- Water trading can facilitate investment and structural adjustment in response to changing conditions (known as 'dynamic efficiency'). For example, in a capped system in which no new entitlements are available, trade enables new water users, such as new 'greenfield' irrigation developments, to establish and develop. The corollary is that water markets provide a mechanism for existing users to retire or move on. As a result, markets enable dynamic changes in the size and composition of water-using industries over time. This is particularly useful in a market-oriented economy such as Australia's, in which farmers face global market forces for the commodities they produce.
- Water trading can also promote 'productive efficiency'. The price signal for water in the market provides an incentive for users to make efficient use of all inputs and invest in improving the efficiency of their on-farm water use.

Trading can make less water available for the environment, as water users aim to use all their now high value allocated water or sell water that they might not have used, which otherwise would have remained in the water resource (Horne 2013). The activation of dormant or partially dormant water rights must be allowed for when introducing water trading.

Case study – Tennessee Valley Authority Reservoir and River Operation Planning, USA

The Tennessee Valley Authority (TVA) manages the 650-mile Tennessee River system to provide multiple and often competing benefits to the 8.5 million people living in the seven-state Tennessee Valley region. The system includes 49 multipurpose dams, which are managed in an integrated manner to reduce flood damage; produce low-cost electricity; enable barges to deliver goods to port; and ensure a supply of water for municipal and industrial use, for cooling TVA's nuclear and fossil plants, and for aquatic habitat. In addition, recreation on the system's scenic rivers and reservoirs generates millions of dollars annually for the region's economy.

A new policy for operating the Tennessee River and reservoir system was introduced in 2004. The policy shifted away from managing specific summer pool elevations on TVA-managed reservoirs to managing the flow of water through the river system. It specifies flow requirements for individual reservoirs to keep below-dam rivers from drying out, as well as the system-wide flow requirements to ensure that enough water flows through the river system to meet downstream needs. This benefits recreation on tributary storage reservoirs while meeting other needs:

> protecting water quality and aquatic resources, ensuring year-round navigation and providing water for power production and municipal and industrial use.
>
> TVA enhances recreational opportunities by restricting the drawdown of tributary storage reservoirs during the summer – from June 1 through Labor Day. During this period, under normal operations, just enough water is released from these reservoirs to meet downstream flow requirements. TVA works to keep the water levels in these reservoirs as close as possible to each reservoir's 'flood guide level' – a guideline that reflects how much storage space each reservoir needs to hold back potential flood waters. When water must be released to meet downstream flow requirements, a fair share of water is drawn from each reservoir (TVA 2013).

Case study – Collaborative management of flood risk: River Tweed catchment, Scotland–England border, UK

The River Tweed catchment straddles the border of Scotland and northern England. Flooding has become an increasingly serious issue with up to 9% of the catchment being considered at risk from flooding, primarily due to river modifications, poor land practices, and floodplain development. This equates to some 4,575 'at risk' properties (Bissett *et al.* 2010). The town of Peebles on upstream River Tweed in Scotland, for example, is now impacted by regular flooding (refer Figure 1.8 in colour plates).

As a result of commitment to the EU Water Framework Directive (WFD) 2000, the Solway Tweed River Basin Management Plan (RBMP) was finalised by the Scottish Environment Protection Agency and the Environment Agency England in December 2009. The EU Floods Directive 2007 requires members to assess and map flood risks by 2013, and prepare flood risk management plans by 2013, in coordination with RBMPs. A strategic aim of the Tweed Catchment Management Plan, updated by the Tweed River Forum in 2010, is to assist in mapping flood risks and 'to adopt a catchment-based approach to flood management to protect the people, property and prosperity of the Tweed catchment whilst respecting its physical, ecological and aesthetic qualities' (Bissett *et al.* 2010: 51). Among other strategies, it identified a series of measures for 'Natural Flood Management' which includes restoration of upland wetlands to increase flood storage, upland forestation, restoring meanders in canalised reaches to slow down flood flows, increase land available for floodplain storage, and increase woody debris in upland rivers to slow down flow rates.

The Tweed Forum has commenced an ambitious program to alleviate downstream flooding with a pilot in the Eddleston sub-catchment. Figure 7.5 (see colour plates) shows intervention with woody debris to slow flow. Re-introducing meanders on floodplains has challenged farmers and long-term grazing patterns however collaboration was enhanced through participatory mapping methods to capture local knowledge and develop a catchment model (refer to Box 4.8 in Chapter 4). The University of Dundee has been involved in research and monitoring to determine the effect of changes to the floodplain (Figure 7.6 in colour plates).

Through a hybridised governance arrangement that includes the non-regulatory EU Directives, various government agencies in both Scotland and England, and the non-governmental Tweed Forum, comprehensive actions have been initiated to manage flooding and assess the effectiveness of actions. The Tweed Forum, similar to catchment authorities or associations elsewhere, is an umbrella organisation that brings together bodies, on both sides of the border, with an interest in the management and welfare of the river and its environs.

8   Comparative assessment

This chapter builds on what has been presented about good water resource planning processes in the previous chapters. The situational analysis (Chapter 5) establishes the basic hydrologic, ecosystem and socio-economic information and identifies risks and opportunities. Objectives are then formulated, together with a logic frame outlining a hierarchy of outcomes that links water resource management to broader benefits and beneficiaries (Chapter 6). Possible actions that could be used to achieve the outcomes are then identified (Chapter 7), which are grouped in management options and evaluated using methods and tools described in this chapter so that a set of actions can be selected.

In accordance with the logic frame for water resource planning (Chapter 3), Chapter 8 therefore demonstrates how to investigate and assess options for actions, using criteria for evaluating options based on principles for good water governance (Chapter 2) and agreed outputs and objectives (Chapter 6). It should be acknowledged that an assessment of options may lead to a re-evaluation of management options and even outputs and objectives, if the range of options do not, after assessment, appear feasible or are unlikely to deliver desired results.

The chapter begins with a brief overview of management option development, which is a frequently used tool in water resource planning to establish a range of options based on varying conditions in an uncertain future.

Water resource models can be used to test how altering management actions changes the water regime characteristics. The bridge between model outputs of water regime changes and the effect on ecosystems and communities of these changes is ecological and socio-economic impact assessment. Techniques and guidelines for impact assessment are presented. These techniques should inform trade-offs with transparency and fairness, mitigate negative impacts (identifying structural adjustment), and minimise conflict. This relies on adequate ecological, economic and social assessment as well as solid stakeholder analysis in the beginning of the process.

Tools and techniques for innovative problem solving and making trade-offs are presented, with a brief discussion on multi-criteria analysis (MCA), consensus building, and the Global Reporting Initiative. MCA aids decision-makers in choosing between options or alternatives with often multiple or conflicting objectives and criteria for evaluation, and in achieving consensus in complex environmental decision problems. This process may involve constructing a structural change package to mitigate negative effects and more equitably share the benefits and disadvantages.

A water resource planner does not need to be an expert in all of these areas, but needs to know when expert assistance is required, what can be produced and how to use it.

Developing management options

In practice many of the actions that are available (or not available) are dictated by government policies, government decisions, legislation or history. For example, a government may have decided to build a dam, and water resource planning is about how much water the dam should be allowed to capture and how it should be operated. Or where a dam is already in place water resource planning is about how it should be operated to meet consumptive needs, environmental flows, and/or requirements such as flood control. Likewise legislation may dictate the nature of water rights and how they are determined and held, and government budgets typically restrict the possible actions that require government funding.

According to Australia's National Water Commission, the greatest challenge in water planning is to find acceptable mechanisms to return extractions to sustainable levels.

> For systems that have not yet reached the need for water recovery, preventive measures are considered less disruptive and more cost-effective. Setting limits and managing extractions within them to minimise the need for water recovery in the future is clearly preferable to the complex, protracted and resource-intensive processes usually required for water recovery. Maintaining extractions within sustainable levels of extraction optimises economic, social and environmental outcomes over the long term (NWC 2013: vii).

A first filter that can be applied to selecting possible actions is simply:

- Given the context, is it feasible to implement?
- Is it likely to support an output and objective?

Feasibility needs to consider not just technical but also social factors and community capacity. Faurès *et al.* (2010: 538) explain that not only is the availability of information important, but the end-users' capacity to act on it. An amazing implication of improved seasonal forecasting for example, is the difference in the capacity between large and small farmers to take advantage of such information, leading in extreme cases to increased production, a reduction in grain price, and further negative impacts on smallholders who cannot benefit from such information as much as commercial farmers.

While many actions are primarily designed to support one objective, others will support multiple objectives: there is not always a one-to-one relationship between objectives (benefits) and actions. For example, actions that facilitate markets in water rights can be used as a means for both maximising economic benefits and redirecting water to ecosystems (as is happening in the Murray-Darling Basin in Australia). Weirs can help support irrigation as well as navigation and recreation. A dam may be designated for electricity generation, water supply, as well as flood control.

Management options often consist of different levels of water availability for consumptive use (implemented through construction and operation of infrastructure, regulation of extraction, or economic instruments) supported by actions to mitigate undesired effects on other objectives. For example, fish ladders can be added to weirs to reduce their impact on fish migration, and establishment and support of markets in water rights can be used to mitigate the economic impact of water shortages or restraining or reducing overall extraction of water for ecosystem benefits (see for example National Water Commission 2010). A plan that reduces water allocations to farmers may require incentives to be provided for training and installation of efficient irrigation or monitoring equipment. At an urban level, it may include subsidies for installation of rainwater tanks, schemes for restrictions and education programs to encourage water saving as occurred in Brisbane Australia during water restrictions in 2006. In the EU, some of these broader aspects are included in River Basin Management Plans and associated implementation measures. In the Ebro Basin Hydrological Plan 2010–2015, for example, a programme of measures aims to achieve environmental targets, meet water demands and adapt to extreme events (floods and droughts) at a cost to water authorities and users of €800 million per year (García-Vera 2013).

Water demand management actions can be applied to both urban and rural areas. In relation to irrigation, it often aims to improve water use efficiency. Brooks (2006: 521) suggests an operational definition of water demand management as:

1 reducing the quantity or quality of water required to accomplish a specific task;
2 adjusting the nature of the task so it can be accomplished with less water or lower quality water;
3 reducing losses in movement from source through use to disposal;
4 shifting time of use to off-peak periods; and
5 increasing the ability of the system to operate during droughts.

This definition brings out the drivers of water saving and permits the tracking of gains by the source of the saving. It is applicable to nations at different stages of economic development. It also shows how goals of greater water use efficiency are linked to those of equity, environmental protection and public participation. Taken together, these goals make water demand management less a set of techniques than a concept of governance.

The identification of suitable actions and grouping of them into management options may be done through an iterative process by a technical group of agency staff and experts, and can benefit from input from a community advisory group. To maximise discovery of the range of possibilities, benefits and impacts, good practice is to include stakeholder input. Including stakeholders in the process can contribute novel solutions, new insights and knowledge, increase robustness of action plans, increase support for policy, and improve legitimacy and quality of decisions. Broader community input can be gained through publishing an issues and options paper at an early stage of plan development and inviting comments and suggestions.

Options can include not only the kind of actions, but also the timing and manner of their implementation. For example, reductions in levels of water allocation can be imposed at a fixed time or phased in gradually over a period of time, as is illustrated in the example in Box 8.1.

Box 8.1: Groundwater allocation reductions in New South Wales, Australia

The extensive alluvial aquifers in the lower Gwydir valley in northwest New South Wales, Australia, were the subject of a multi-year water resource plan development process in the early 2000s. Groundwater extraction for irrigation and commercial purposes had been subject to licensing for many years, and the government stopped issuing licences in 1993. Despite this, studies had demonstrated that total water extraction was exceeding sustainable levels and there was still large potential for increased extraction under issued water licences.

After extensive research, modelling and community consultation, an annual average sustainable level of extraction of 32,300 ML/year was agreed to. Average extraction in the years 1993–94 to 2000–01 was over 32,000 ML/year and trending upward. With total licensed entitlements in the system of 68,000 ML/year, potential for increased water extraction was large. A reduction in licensed entitlement to water was required to provide for ongoing sustainability. However, the way that this was to be implemented became a major sticking point.

Initially, the state government determined that reductions would be applied proportionally to all water licences, on the basis that the licences had developed a value reflected in the much larger value of properties with water licences. However, this created considerable community angst, as many licence holders had fully developed their licences and would be hit much harder than those who had not developed or had only partially developed. After extensive lobbying the government's position was altered so that the reduction would be distributed to licences based on a formula that took account of both the specified volumetric water entitlement and the extent that the entitlement had actually been used.

To help to mitigate the impact, the state government agreed to implement the reduction stepwise over a ten-year period, and provide (with the federal government) financial 'adjustment' funding to affected water users. Additionally, the water resource plan provided for temporary and permanent trading of water licences, and for 'carryover' of annual allocations so that water allocations could be banked from one year to the next.

See Hamstead *et al.* 2008: Appendix E, and Barrett 2009 for further information.

Comparative assessment

As part of the water resource planning process, management actions are to be selected that achieve to the extent possible the defined outputs and objectives. Commonly trade-offs need to be made because objectives are to an extent mutually exclusive. Sometimes these trade-offs are black and white but mostly they are not easy or clear. Consider for example a wetland. It is clear that if water is totally cut off that it will die. But if watering events are reduced to provide more water for irrigation, the response is less certain. Scientific assessments can attempt to quantify the risk, but there is always a level of uncertainty. Often what is being traded-off is a level of risk. For example, a certain watering regime may be assessed as imposing a moderate risk to the environmental values of a wetland. The trade-off is not the abandonment of the wetland, but the exposure of it to greater risk.

Overall, one assesses management options through *comparative assessment*, considering whether and how each option would contribute to, impact on, or change the risk to, objectives. Ideally these are stated explicitly in the plan (as described in Chapter 6) but where they are not, they may be in broader legislation or agreements that overarch the plan.

The kinds of objectives that occur in water resource plans can be encompassed under the three goals of IWRM discussed in Chapter 1:

- economic efficiency – to make scarce water resources go as far as possible and to allocate water strategically to different economic sectors and uses;
- social equity – to ensure equitable access to water and to the benefits from water use, between women and men, rich people and poor, across different social and economic groups both within and across countries;

- environmental sustainability – to protect the water resources base and related aquatic ecosystems and more broadly to help address global environmental issues (Jønch-Clausen and Fugl 2001).

To assess whether these objectives are met, tools to estimate benefits, risks and impacts can be considered. They are commonly grouped into socio-economic assessment and ecological (or environmental) assessment tools. Other tools can also be applied: Strengths-weaknesses-opportunities-threats (SWOT), Global Reporting Initiative for sustainability and governance, and multiple criteria analysis (MCA). We consider these later in this chapter.

Sometimes a bottom line or minimum requirement must be considered. The Australian National Water Initiative (Commonwealth of Australia 2004) indicates that in making water resource plans, state and territory government jurisdictions should optimise economic, social and environmental outcomes, provided that extraction of water is within environmentally sustainable levels (clauses 2 and 23). Applying such concepts on the ground has proved challenging. Hamstead (2009: 1) comments:

> The National Water Initiative requires the return of overallocated and overused systems to environmentally sustainable levels of extraction (cl 23). While such terms as 'environmentally sustainable levels of extraction' and 'overallocation' are defined in the NWI, each Australian jurisdiction has different views on whether particular systems are overallocated or not and has taken different approaches to determining the level of stress or risk environmental values are exposed to when weighed against economic impacts. For many parts of Australia, this manifests itself as the debate about how much additional water can be taken from rivers or aquifers without there being an unacceptable risk of environmental damage. For water systems where extraction levels are already high and there is strong evidence of environmental decline, modifying or reducing extraction to halt environmental loss has proven to be one of the most difficult aspects of Australia's water reforms.

Uncertainty about the future provides a particular challenge for assessing and comparing management options. One might ask which will perform best in a range of possible futures? For this reason we consider dealing with uncertainty as a particular aspect of comparative assessment.

Defining social equity is often difficult, as it is highly dependent on the perspective and perceptions of persons or groups. For example, which is more important, economic benefits in a town or in a rural area? Achieving this kind of objective is as much about process as it is about outcomes. In many cases an overriding issue in decision-making, which is rarely explicit, is perception of equity and fairness. Lack of equity provides fuel to conflict. From a review of case studies around Australia, Hamstead *et al.* (2008) suggest that significant unaddressed concerns of a particular stakeholder group are likely to result in change to a plan because that group will continue to use all the political and legal processes available to them to have their concerns addressed. We therefore discuss equity and fairness as an important aspect of decision-making.

Dealing with uncertainty

A major component of water resource management is about managing for the uncertainty related to matters such as incomplete data and climate variability. Consideration can be given to either a) reducing the causes of uncertainty or b) mitigating the potential consequences. Water resource planning processes need to address uncertainty in order to achieve its objectives, particularly that of water security.

Agriculture is one of the most vulnerable industries to climate variability and the unpredictability of precipitation, temperatures, wind, and associated river runoff, and groundwater recharge. Farmers have long accommodated routine risk through risk-mitigation strategies and a cautionary approach. Predictability is an important factor. Not only the amount of water but timing, volume, rate of supply, lack of guarantees or reliability of water delivery are among factors such as market conditions that farmers need to take into account when deciding on crop management. Strategies can vary from aiming for a high possible income, or producing a low but more reliable income, or to simply aim for a regular food supply (Faurès *et al.* 2010). It often results in a conservative attitude, selecting 'low-value but more resistant crops that can sustain relatively long periods without water supply' (Faurès *et al.* 2010: 533).

Storing water in large system-wide reservoirs or farm dams as well as underground with accompanying irrigation systems have mitigated risk in some areas, reducing uncertainty and compensating for variable climate and unreliable supply. However improvement in irrigation efficiency and adapting farming practices can also assist such as: closed irrigation channels; laser-levelling of fields and drip or other efficient systems to promote more even distribution of irrigation water; and computerised farm-level crop monitoring.

As discussed in Chapter 5, in water resource planning, scenario development can be used as a decision-support tool to explore possible futures; addressing uncertainty without committing to a particular forecast. Scenarios are a means of testing the plausible outcomes of the implementation of management options in an uncertain future. While a variety of qualitative (e.g. brainstorming, narratives) and/or quantitative (e.g. computer modelling) methods and tools can be used, in water resource planning, hydrologic models usually form the basis for scenarios which are then used to estimate impacts on environmental, social and/or economic factors.

Management options can be input into models that include a range of future climate and development scenarios to allow their outcomes to be compared in terms of robustness to an uncertain future. Where there is uncertainty in the relationship between water regime characteristics and the achievement of economic, social and environmental objectives, a range of cause and effect scenarios can be added. This can potentially create a very complex array of outputs. For this reason the range of future climate/development scenarios and cause/effect scenarios has to be limited.

Variability in weather has always been a constraint to farming, but by 'increasing uncertainty, climate change will only add a new dimension to a very old challenge' (Faurès *et al.* 2010: 540). Importantly it reinforces the need for a precautionary approach, and effective models that are open, transparent and verifiable.

Equity and fairness

In Chapter two, it was suggested that principles for water resource planning need to have fair, participatory, transparent and accountable processes and mechanisms to equitably allocate the benefits of water. These are generally referred to as procedural and distributional equity. Lukasiewicz (2012) has added an additional element, interactional justice, to form a social justice framework. These are discussed in turn.

Fairness in outcomes, that is, the way the benefits or costs are shared, is known as *distributional fairness*. It generally refers to equitable sharing of the benefits of the resources, not necessarily equal amounts of water shared. This enables aspects such as compensation to be taken into account.

Equitable sharing of the costs of ecosystem maintenance and the benefits of ecosystem services is a fundamental, yet often overlooked, aspect of environmentally sustainable management. Major international conventions (Ramsar, Agenda 21, Convention on Biodiversity) all recognise that addressing the current needs of people in an equitable manner is essential. Australia's National Strategy for Ecologically Sustainable Development has as one of its three objectives to 'provide for equity within and between generations'. The precautionary principle is an attempt at a practical manifestation of intergenerational equity. It implies that 'serious or irreversible damage' fails the intergenerational equity test, and that the threat or risk of this needs to be managed, and indeed prevented. Again, much discussion continues among academia and practitioners about the precautionary principle, and its related concepts of 'serious' or 'irreversible' damage.

Equity and fairness is achieved largely through the way the decision-making process is undertaken. This includes using a process that is seen to be fair and transparent, allowing for consideration of all views – *procedural fairness*. Syme *et al.* (1999) provides an excellent analysis of fairness in water resource planning in Australia based on a series of case studies over a decade.

Procedural fairness contributes to, but does not guarantee achieving distributional equity. Additional practices that are important for this, are:

- Inclusion and influence in decision-making by those affected by a decision and those interested in the decision;
- Transparency, accountability and openness with sharing of information as well as justification of decisions made;
- Neutrality in decision-making;
- Identification and valuation of the services provided by ecosystems and to identified beneficiaries;
- A willingness to identify and consider a broad range of management options to achieve the desired outcomes, including giving affected parties an opportunity to identify such options;
- A robust socio-economic assessment of management options, which identifies risks to potentially affected parties and the mechanisms that are available to address those risks;
- A process for weighing up the value of the services and the costs of maintaining the ecosystem attributes, which takes account of non-market public good services that are difficult to value in economic terms;
- A recognition that public funding may be needed for equitable sharing of costs to occur (often referred to as benefit sharing).

These principles should be included in the earlier steps in water resource planning described in previous topics, and reflect the importance of the integrity of the whole planning process in achieving equity, rather than just the decision-making at the end. If a process for identifying and valuing assets and functions has not met broader public expectations it will soon become evident.

The third aspect, *interactional or interactive justice* refers to interactions between decision-makers and stakeholders, at a more personal level that embed trust, respect, honesty, recognition, legitimacy (of decision-maker and stakeholders) and propriety (Lukasiewicz 2012).

Tools and techniques for assessing options

Use of hydrological models

Hydrologic/hydrogeological models are the basis for option development and assessment of impacts. Development of some kind of model is normally done as part of situational analysis. As discussed in Chapter 5, these models can vary from simple to complex, depending on the level of competition for water and available knowledge. In the situational analysis step the models provide baseline information on system behaviour and projections of future inflow/recharge patterns and how they might affect the water regime.

In this step the models can be used to test how different management options change the benefits and impacts to ecosystems and water available for extraction. The models thus need to be able to simulate how any proposed management changes will affect the flow regime. For example, if new infrastructure is being considered, the model must include capacity to simulate what would happen to flows with such infrastructure in place. The planner generally asks the modeller 'what if ?' questions, and expects that the modeller will provide a response in a form that is useful. The planner needs to understand broadly the kind of outputs that models can produce, and the limitations of the model. Examples of model outputs are shown below.

For the Tasmanian River Clyde Water Management Plan 2005, a river system model was used to simulate different environmental flow rules and estimate how they might affect water available for extraction. Table 8.1 compares the amount of water that would be available during the period May to October, with and without the Environmental Water Provisions in place. Table 8.2 shows how the cease to take rule might affect the volume available to be taken and the days on which water can be taken during the winter months when the majority of water is taken into on farm storages. Three different time periods are used to calculate the availability to show the impacts of the changes in the system due to the recent drought years and increases in storage development in the catchment. The three periods are the whole period of record 1979 to 2003 to reflect the long term range; the period between 1988 and 2003 which is when significant storage developments were in place in the catchment and the initial stages of the drought period; and the period between 1998 and 2003 to reflect the driest period of the record. In this case the model did not attempt to forecast future flows, but rather used historic flows and left it to the planners and decision makers to use this information as a basis for estimating future impacts.

Table 8.3 shows how a model was used to estimate the effect of a change in system reserve policy would affect allocations in the Goulburn regulated river system in Victoria. In this case the model compared the effect under three future inflow scenarios, the base case being a continuation of past average conditions, scenario B being the climate that is projected for 2055 assuming a medium level of climate change, and scenario D being the continuation of the recent low inflow patterns. It can be seen that the proposed policy has minimal effect on average annual diversions, but trades off a reduced number of years of very low allocations for a reduction in years with higher allocations.

Assuming that some sort of model is available, consideration is now given to methods for relating changes in the water regime to effects on ecosystems, social, cultural and economic values.

Ecological impact assessment

This involves the estimation of how projected changes in the water regime for different options will affect the integrity and viability of ecosystems. It is thus the bridge between model outputs of hydrology changes and the effect on ecosystems of these changes. In Chapter 5 we discussed methods that are used to identify and value ecological assets and processes and determine

Table 8.1 Impact of proposed environmental flows on water availability, River Clyde Water Management Plan 2005, Tasmania, Australia

Reliability	*Water Availability – ML per winter*					
	1979–2003		*1988–2003*		*1998–2003*	
	With eflows	*Without eflows*	*With eflows*	*Without eflows*	*With eflows*	*Without eflows*
Maximum	98,563	103,632	38,308	44,841	38,308	44,841
3 years in 10	20,464	27,927	16,773	20,500	7,290	9,383
5 years in 10	12,976	16,889	10,554	14,124	4,206	5,806
8 years in 10	3,969	5,801	4,127	5,804	2,800	5,034
10 years in 10	560	2,208	560	2,208	560	2,208

(Source: Adapted from DPIWE 2005: 31. Used with permission.)

Table 8.2 Impact of cease-to-take rule on access to water River Clyde Water Management Plan 2005, Tasmania, Australia

Year	*Without cease-to-take rule*		*With cease-to-take rule*		*Reduction in water available due to rule*	
	Water available (ML)	*No. of days cannot take*	*Water available (ML)*	*No. of days cannot take*	*ML*	*Days*
1979	10,461	1	7,139	35	3,322	34
1980	15,374	0	11,629	7	3,745	7
1981	51,878	10	48,328	31	3,550	21
1982	2,374	60	973	133	1,400	73
1983	33,764	27	30,398	35	3,365	8
1984	57,942	50	55,445	91	2,497	41
1985	28,128	0	24,387	9	3,741	9
1986	103,632	0	99,857	0	3,775	0
1987	3,252	12	1,101	139	2,151	127
1988	28,494	7	25,001	40	3,493	33
1989	18,405	14	15,188	51	3,217	37
1990	21,024	33	17,933	58	3,091	25
1991	13,634	50	11,418	113	2,216	63
1992	27,457	28	24,261	50	3,196	22
1993	14,124	0	10,510	31	3,615	31
1994						
1995			No data available			
1996						
1997	20,151	3	16,576	16	3,575	13
1998	5,798	49	3,812	125	1,986	76
1999	2,208	76	560	153	1,648	77
2000	5,034	123	3,964	152	1,070	29
2001	44,841	38	41,764	56	3,077	18
2002	5,813	107	4,600	140	1,214	33
2003	12,954	64	11,274	70	1,679	6
Average	23,943	34	21,187	70	2,756	36
Median	16,889	28	13,408	54	3,143	30

(Source: adapted from DPIWE 2005: 33. Used with permission.)

Table 8.3   Impact on reliability of supply from proposed change to the seasonal allocation
policy in the Goulburn system

Option	Indicator	Scenario		
		Base case	Scenario B	Scenario D
Current reserve policy	0% Aug. allocation	0 years out of 100	2 years out of 100	11 years out of 100
	< 5% Aug. allocation	0 years out of 100	4 years out of 100	18 years out of 100
	Min Feb. allocation	27%	0%	0%
	< 30% Feb. allocation	1 year out of 100	4 years out of 100	9 years out of 100
	100% Feb. allocation	96 years out of 100	79 years out of 100	28 years out of 100
	Av. annual diversion (GL)	1,638	1,389	1,139
New reserve policy	0% Aug. allocation	0 years out of 100	0 years out of 100	0 years out of 100
	< 5% Aug. allocation	0 years out of 100	0 years out of 100	2 years out of 100
	Min. Feb. allocation	35%	20%	10%
	< 30% Feb. allocation	N/A	1 year out of 100	4 years out of 100
	100% Feb. allocation	93 years out of 100	69 years out of 100	25 years out of 100
	Av. annual diversion (GL)	1,635	1,386	1,128

(Source: Victorian Government 2009: 90. Used with permission.)

the relationship between aspects of the water regime and those assets and
processes. In the comparative assessment step in the water planning process,
this information is used to estimate the impacts of the changes to the water
regime arising from proposed management actions on those assets and
processes. Some examples of approaches are shown below.

Figure 8.1 (see colour plates) shows a sample of a diagram used in the
assessment of options in the development of the Burnett Basin Water Resource
Plan in Queensland. In this case the colours (red – yellow – green) on the charts
represent the risk of impact on environmental values associated with changes to
different flow statistics. These were prepared by environmental scientists, based
on estimation of the cause-effect relationship between flows and ecological
condition. They are thus the bridge between the flow information generated
by models and the likely effect on ecosystems. The model generated the flow
statistics that would apply for different allowed water use options, and these are
shown as symbols on the charts. This proved quite popular with stakeholders
as a way of explaining risk associated with different flows.

In the Northern Sustainable Water Strategy in Victoria, environmental studies developed linkages between the provision of components of flows and the environmental outcomes in rivers and wetlands as shown in Tables 8.4 and 8.5. The actual flow levels for the components were identified for each river.

Based on these flow components, the needed additional average flow volumes to achieve them in each major river system were assessed using models, as shown in Table 8.6.

Table 8.4   Environmental outcomes for a river reach and required flow components

Category	Objective/Outcome	Flow component
1	Protection of drought refuge	Baseflows
2	Protection of drought refuge plus dry spell breaking	Summer minimums throughout the year and every third year deliver winter minimums and freshes
3	Sustainable population of priority instream species	All summer and winter minimums and freshes at recommended frequency
4	Healthy instream environment	Category 3 plus bankfull flows
5	Healthy instream environment and protection of priority wetlands	Category 4 plus reduced overbank flows (one in every three years)
6	Full environmental flows	All recommended environmental flow components

(Source: Victorian Government 2008b: 96. Used with permission.)

Table 8.5   Environmental outcomes for wetland systems (which can be supplied under regulated flow conditions) and required flow conditions

Category	Objective	Flow conditions
1	Drought refuge, basic habitat maintenance and survival of biota for future recolonisation at priority sites	Priority sites managed at dry spell tolerance
2	Drought refuge at all sites	All sites managed at dry spell tolerance
3	Healthy, sustainable, breeding populations and high quality habitat maintained at priority sites. Drought refuges maintained at all sites	Priority sites managed at recommended flood frequency. Remainder of sites managed at dry spell tolerance
4	Healthy, sustainable, breeding populations and high quality habitat maintained at all sites	All sites managed at recommended flood frequency
5	Healthy, sustainable, breeding populations and high quality habitat maintained at all sites. Targeted overbank flooding in small proportion of sites	All sites managed at recommended flood frequency. Targeted overbank flooding
6	All wetland and floodplain assets healthy. Rapid increase in population growth, opportunities for population dispersal and mixing	Extensive overbank flooding

(Source: Victorian Government 2008b: 97. Used with permission.)

Table 8.6   Additional volumes (GL/year) needed for different environmental outcomes under different future flow scenarios, Northern Victoria, Australia

Category of environmental outcome	Ovens River	Broken Creek	Broken River	Goulburn River	Campaspe River	Loddon River
Base case						
1	0	N/A	0	0	0	0
2	0	N/A	N/A	23	4	4
3	0	N/A	N/A	84	12	12
4	0	25	N/A	98	18	25
5	0	N/A	N/A	250	N/A	30
6	3.9	N/A	N/A	TBD	30	44
Scenario B						
1	0	N/A	0	0	0	0
2	0	N/A	N/A	46	6	4.5
3	1.9	N/A	N/A	97	19	15
4	6.7	N/A	N/A	TBD	N/A	31
5	13	N/A	N/A	TBD	N/A	38
6	N/A	N/A	N/A	TBD	N/A	N/A
Scenario D						
1	0	N/A	0	0	0	0
2	0	N/A	N/A	61	9.7	5.8
3	2.8	N/A	N/A	128	31	19
4	9	N/A	N/A	TBD	N/A	41
5	20	N/A	N/A	TBD	N/A	48
6	N/A	N/A	N/A	TBD	N/A	N/A

N/A – not applicable; TBD – to be determined

(Source: Victorian Government 2008b: 99. Used with permission.)

Socio-economic impact assessment

Socio-economic assessments relate to how the changes to the management of water are likely to impact people, communities and economies. As with ecological impact assessment, social and economic assessments draw on the information on the relationship between water and social and economic benefits developed in the situational analysis.

Social assessments should include 'cultural impacts involving changes to the norms, values, and beliefs' (Burdge 2004: 3) and identify how affected people will respond in attitude and actions. Perceptions and attitudes are important variables that can lead to real consequences, such as land specu-lation or fear of losing one's livelihood. As a result, social science methods (such as surveys and interviews) are used, supplemented with public involvement procedures and consultation with the affected population (ibid.: 3). Variables include potential changes and impacts on: population compo-sition and distribution; community and institutional structures; political and social resources (including power and conflict); individuals and family (including stability, networks, social well-being); and community resources such as community infrastructure, land use patterns, and cultural resources (Vanclay *et al*. 2004).

Economic assessment typically indicates the change in the value of production and employment associated with the use of water and also includes indicators of indirect effect on other businesses reliant on the water industry.

In Australia, the NWI suggests that social and economic assessments should contribute to water planning by:

- advising on possible social and economic impacts of proposals and options, both positive and negative, which are used to inform trade-offs and which are seen to be transparent and fair;
- identifying ways of mitigating negative impacts, including structural adjustment and flexible and innovative systems;
- providing accurate information to all stakeholders on which to base decisions, which functions to engender trust and support for the process;
- providing an understanding of the cultural context in order to: identify public benefit outcomes, take relevant values into account (including Indigenous social spiritual and customary objectives and strategies and possible Native Title rights), and minimise conflict.

Assessment of socio-economic impacts can be considered to cover:

- identification and analysis of possible impacts
- identifying responses to impacts
- developing a mitigation plan.

Identification and analysis of possible impacts should be done in relation to each option, and could include predicting conditions with and without the option. Investigation of probable impacts involves the following sources of information:

- data from the agency or proponent;
- census and other statistics;
- record of experience and documents or other literature regarding similar actions;
- field research, including informant interviews, hearings, group meetings, surveys of general population, targeting in particular those who may wear disproportionate impacts.

There are several types of analyses that can be used including: comparative statistics, cost benefit analysis, option modelling, expert judgment, and opportunity cost or future foregone. Plant *et al*. (2012) documents a range of economic valuation techniques for benefits and services associated with water resources that can be applied in different circumstances. Secondary and cumulative impacts as well as alternatives should be identified (IAIA 2003). Preferably impacts of each management option under a range of future scenarios are considered to address uncertainty.

Identifying responses to impacts includes identifying how affected people will respond in attitude and actions. This could be done by comparing impacts with responses of other similar communities to change; however, it is most effective as a result of surveys, interviews and public consultation. Surveys can investigate effects on amenity, community well-being, social cohesion, cumulative change and ability of the community to respond to change. Risk and vulnerability assessments done in the situational analysis, as discussed in Chapter 5, can identify where impacts could potentially 'blow out' due to high sensitivity to change.

Mitigation measures – A socio-economic impact assessment should not only forecast impacts, it should identify any available means to mitigate adverse impacts. As part of a typical assessment, a mitigation plan is frequently developed to identify the means to mitigate adverse impacts. Mitigation includes:

- avoiding the impact by not taking an action;
- minimising or reducing impacts through redesign, sequencing or operation of the project or policy;
- compensating for irreversible impacts by providing substitute policies, facilities, resources, or opportunities. This could include direct compensation or enhancing other quality of life variables as compensation and ensuring that proponents and leaders are sensitive to the feelings of the community (IAIA 2003).

Role of committee members and the community

The most common form of identifying socio-economic values and impacts either explicitly or inadvertently by jurisdictions has been through input from community committees and other public consultation such as public meetings and submissions. However, relying solely on community input does not necessarily ensure transparency, thoroughness, or accuracy in either identifying or addressing impacts. This is not intended to undervalue the input from the community through the planning process as it is absolutely necessary; however, such input relies on effective community engagement and skilled facilitators, transparent documentation of issues and values, and a commitment and expertise by government to effectively respond to issues. If a range of management options have not been identified or implications assessed, then a negative response to a proposal provides few tested alternatives.

Independent socio-economic or cultural studies can contribute factual information that can facilitate deeper understanding and put competing values in perspective. They can be used to take the 'heat' out of a situation, providing an objective assessment. In the Burnett water resource planning process, such an assessment put in perspective the future needs for urban growth compared to agriculture (Hausler and Fenton 2000) In some cases, they reveal the values of constituents of which their representatives were not aware. Such studies should include primary data collection from the relevant community and region in a methodologically sound manner. Community input is appropriate to identify perceived gaps or raise doubts about the findings of a governmental socio-economic assessment study.

Guidelines on how to do socio-economic assessments

One readily available guideline is that produced in New South Wales, Australia for use by community advisory committees in developing their first round of water sharing plans. An expert Independent Advisory Committee on Socio-economic Analysis (IACSEA) produced *Socio-economic Assessment Guidelines for River, Groundwater and Water Management Committees* (IACSEA 1998).

These guidelines differentiate between a 'profile' (usually done in the situational analysis stage) and 'impacts'. They include both as part of a 10-step process. The guidelines clearly identify the outcomes of each step, accompanied by a quality assurance standard. They also describe methods and proformas that can be used to undertake the work. An important step in the process is 'identifying all effects of water management options' from a number of perspectives, including effects on:

- both extractive and non-extractive uses of water
- different population groups, e.g. low income, Aboriginal communities, aged
- different industry sectors
- different communities, e.g. geographical, occupational
- other effects – over time, e.g. short vs. long term; locations e.g. local and regional, upstream and downstream.

This step assists in understanding the nature of the effects of each option and how they are distributed across the range of water uses. The next step 'assessing effects' identifies the extent, likelihood, intensity, timing and duration of the effect. A preliminary assessment enables screening options for further consideration and evaluation if necessary; some options may be eliminated at this point. A more detailed assessment provides a clear and transparent account of effects, including the most affected stakeholders and their concerns.

Another step, 'identifying the preferred option' relies on ranking alternatives based on criteria which reflect:

- desired outcomes and objectives of the plan as well as agreed principles
- social, economic, financial and environmental effects of each option

- time and cost of implementing the options
- possibility of implementing the option under current legislation and policies.

This information may be fed into other decision support processes discussed below such as multiple criteria evaluation.

Box 8.2: Socio-economic impact assessment, Coopers Creek, Australia

The IACSEA methodology was applied in determining socio-economic impacts of recommended change to the flow rules for the Coopers Creek Water Sharing Plan (WSP) in New South Wales in 2009 (Singh *et al.* 2009). It assessed the economic impacts of a change in the current Cease to Pump (CTP) from its original low flow value of 20 ML/d in July, August and September (JAS) and 14 ML/d in October to June (OJ) to:

- Option 1 – an out of court settlement to raise the value to 31 ML/d in JAS and 17 ML/d in OJ; and
- Option 2 – lowering to 10 ML/d in all months from its current value of 20 ML/d in JAS and 14 ML/d in OJ.

As lowering the CTP levels would allow more opportunities to extract water and therefore benefit irrigators, there was no further analysis undertaken for option 2. To assess impacts of option 1, in the absence of Australian Bureau of Statistics Agriculture Census 2006 data at an appropriate scale, the New South Wales Department of Water and Energy (DWE) undertook a survey of all 92 water licence holders in June 2008, achieving a 65% response rate. The DWE also analysed the impact of changed hydrology on cropped land and irrigation water crop requirements. As the unregulated river system was not metered, data on monthly irrigation requirements were estimated through the licensee survey.

Option 1 was found to have significant negative economic impacts for some irrigators, mainly in the dairy industry. Over the 114 year period the impact would be $ AUD 7.1 million; at an average annual impact of $ AUD 62,000 in the region. This represents 5% of the gross value of irrigated production in the region. The effect would vary according to the year, and there was a concern was that smaller operations would not have sufficient reserve to survive economic pressures during low flow years.

Irrigators were also asked how the proposed changes to CTP levels affect the way they manage water use. About 33% would not change the way they have been using water; 28–32% would apply less water/ha in the two different time periods; 16% and 8% would reduce the area irrigated under the two different time periods.

The survey results indicated that a large number of irrigators would need to develop their management strategies to mitigate the resulting impacts. The WSP amendment took this into account, allowing five years before the Option 1 changes take effect, giving some time for transition. Further assistance might be required from off-farm sources including government and relevant industry sources. These are aspects not usually covered by New South Wales water resource plans. This study did not compare impact on consumptive uses with non-consumptive values of the system. Given available resources and a decision to meet environmental flows, it was decided to focus on those most negatively impacted. As indicated in the Guidelines, it can be quite appropriate to screen and focus assessment on certain options.

Indigenous values and cultural studies

As mentioned in chapter 5, specific applied research may be needed to identify values, interests, connections and relationships of water with specific groups of people at the appropriate scale. This kind of information can be used both to identify potential impacts of management options on values, and to tune management options to better achieve them. We provide a study identifying Indigenous values in northern Australia as an example in Box 8.3.

Box 8.3: Indigenous values in the Roper River, NT, Australia

In the case of the Roper River in the Northern Territory in Australia, statutory and common law land claims have resulted in very high rates of Indigenous land ownership and therefore relatively stronger bargaining power. 70% of the regional population are Indigenous and they hold title to approximately 70% of the land. Indigenous land supports a variety of uses, including areas for community living and cattle grazing.

Archival and ethnographic material were used to identify the significance of water to Indigenous people with 18 people interviewed, selected on the basis of seniority, group identity, knowledge of the country, place and duration of residence, recent profile in speaking about water issues, and expected availability for interview.

Among other aspects, significant permanent water holes and springs, water as part of the larger landscape, as well as protocols and practices governing cultural water sites and landscapes were identified. They saw their role not solely as guardians of country, but also as potential beneficiaries of existing and future productive activities. This is based on their assertions of ownership, their desire for recognition; and intent to mitigate their economic marginalisation.

The draft Mataranka (Tindal Aquifer) Water Allocation Plan 2011 included an outcome to enhance the benefits to Indigenous people 'through protection of culturally significant water-dependent sites as well as providing access to water for commercial development' (DNRETAS 2011: 11). The plan committed to a redistributive mechanism which reserved a substantial volume of water for Indigenous people to use for economic purposes, called the Strategic Indigenous Reserve. This amounted to 25% of the commercial allocation, which ensured that as 'late entrants to a water market with economic aspirations that are still taking shape, Indigenous people could be assured of access to water for commercial purposes' (Jackson and Barber 2013: 12).

The delayed establishment of the plan for three years as a result of a change of government illustrates the 'political risks faced by Indigenous people in trying to gain a share of the resource' (ibid: 13).

Multi-criteria analysis

The term multi-criteria decision-making is most often used to refer to techniques used to aid the decision-maker or decision-makers in choosing between options or alternatives with often multiple or conflicting objectives and criteria for evaluation. It has been particularly successful in aiding understanding and meeting consensus in complex environmental decision problems. It provides a structure for complex decision-making problems that allows a problem to be broken down into workable units, in such a way that the complexities of the problem are transparently unravelled. This is essentially done through an interactive process of identifying options, criteria and preferences. Weighting of criteria allows the exploration of the consequences of giving priority to specific experts and/or water user interests (Barton *et al.* 2010). Such methods can be used to identify a single most preferred option, to rank options, to shortlist a limited number of options for subsequent detailed review, or simply to distinguish acceptable from unacceptable possibilities (Proctor 2010). In contrast to benefit-cost analysis it does not require all impacts to be expressed in monetary units, or even quantitatively (Hajkowicz *et al.* 2000).

Multi-criteria analysis (MCA) refers to decision-aiding techniques that:

- help in identifying key issues or problems
- decide between a finite number of options to address these issues

- use a set of criteria to judge the options
- assign weights to each objective/criterion
- use a method by which the options are ranked, based on how well they satisfy each of the criteria (Proctor 2009).

Features of an MCA include:

- simplification of complex decision problems using an interactive decision support software;
- providing an important means of communication and structure for discussion between the decision making team and the wider community;
- a diversity of evaluation techniques that enable use of qualitative and quantitative data;
- decision makers can specify the level of complexity: it can be simple or elaborate to suit the application and needs of decision-makers;
- the ability to identify trade-offs;
- the imposition of structure and transparency on the problem by identifying options criteria and preferences for criteria;
- sensitivity analysis of results (to determine significance of individual data values in determining the final result and can take account of uncertainty in estimating figures involved) (Proctor 2009, Hajkowicz *et al.* 2000).

One of the limitations of the MCA is that some techniques are extremely complex, must be undertaken by a specialist, and as such may detract from transparency or even the decision-maker understanding the trade-offs. On the other hand, graphical illustrations of implications can be used to enhance transparency through including stakeholder input and judgements in the decision-making. A variety of MCA techniques has been used in different contexts, making MCA approaches difficult to compare. There is little guidance as to which should be used in different circumstances. However, becoming familiar with techniques used in the different cases presented below can foster insight and deliberation into how it could be used or adapted for other purposes.

One example of its use in water resource planning in Australia is documented in Hamstead *et al.* (2008) in relation to the Victorian Central Region Sustainable Water Strategy. In this case a multiple-criteria 'sustainability assessment' of options provided a method of integrating a wide range of factors into the decision-making process. Before being short-listed, options were measured against five key assessment criteria of: realistic, scale, impact (environmental), cost and fairness. The short-listed options were then assessed against environmental, economic and social criteria (including social acceptability and fairness).

Assessment of each option was presented in tables as an appendix to the draft strategy (see Figures 8.2 and 8.3 in colour plates for sample information). The results are colour-coded in shades ranging from dark orange to dark blue. A dark orange square indicates an option has a highly negative score. A lighter orange square still indicates a negative score; however, the costs or impacts may not be quite as severe. Similarly, a dark blue square indicates a highly positive score. A light blue square would also indicate a positive score, although again the benefits may not be quite as pronounced. A yellow square indicates a neutral score.

A key aspect of MCAs is how the different aspects are weighted against each other. One process for coming up with weightings is to facilitate an iterative deliberative process (Deliberative Multi-Criteria Evaluation – DMCE) among the key stakeholders that enables participants to discuss and agree, or approach agreement, on weightings applied to criteria and consideration of scenarios and/or choices (Proctor 2009). Failure to adequately engage members of the community impacted by the decision may result in overlooking criteria or alternatives of importance (Hajkowicz *et al.* 2000). Salgado *et al.* (2009) combined social research techniques and participatory MCA to reveal framings, values and interests of stakeholders. The process helped explore previously unforeseen alternatives, was useful for problem structuring in a collective, flexible and iterative way and improved the information exchange and reflection process.

In 2009, a DMCE process was carried out in the Howard River catchment in the Northern Territory, Australia, in a region where there is increasing pressure on the quantity and quality of ground and surface water resources (Proctor 2010). Residential and agricultural development in Darwin's hinterland has increased competition for groundwater from the bore field that supplements the Darwin metropolitan water supply, thus generating tensions between different water users. In this case, the DMCE process was used to assist in revealing and settling trade-offs between competing outcomes in a transparent way.

In Norway, MCA contributed to a review of regulations for the Glomma River Basin. With 56 hydropower stations and 26 hydropower reservoirs, hydropower is a major use. In the past, a loosely defined expert panel assessed minimum flows, compensation flows, or environmentally motivated rules for water-level variations in reservoirs. In this case, an MCA was used to improve the documentation of expert judgement in the establishment of environmental flow criteria and explore trade-offs between hydropower and other benefits (Barton *et al.* 2010). For the purpose of their study, 'environmental flows' was defined as

> Adopting water release manoeuvring rules for the different reservoirs to obtain as favourable water levels (and water flows) as possible for the total river ecology and the human water use interests, within the constraints set by the economic feasibility of the regulation. This applies both for reservoirs and river stretches (Berge *et al.* 2010: 109).

As indicated by this definition, the approach was to evaluate the value of a range of river-reliant human and ecological benefits and compare them to the economic losses (or gains) associated with hydropower, for different reservoir release rules. In the original study, ecological values were based on water needs for fish, two types of water birds, and macrophytes. 'User' values were boating, farming (irrigation) and hydropower.

In simple terms, the method involved developing a model of the relationship between water levels in the river and each of these values ('pressure-impact curves'), then seeing how they are each affected in different flow scenarios. The MCA approach then quantifies and brings together the change of each value for each reservoir release scenario into a single overall value change for each scenario, allowing the scenarios to be compared.

In this case the method consisted of the following steps (Berge *et al.* 2010: 116; Barton *et al.* 2010: 37–38). In a typical water planning process, the first two steps would have been completed at the situational analysis stage.

1   Identify key river ecological values, e.g. fish, river bed fauna, waterfowl, or proxies.
2   Identify key water use values, e.g. hydropower, domestic water supply, irrigation, fishing.
3   Form an expert panel of professional experts and local water users experienced in values above, e.g. water authorities, scientists, community stakeholders.
4   Expert panel identifies ecological and user interests for which impacts are evaluated.
5   Identify critical time periods, locations (e.g. migration or spawning periods, sailing depth in boating season) and optimal water level curves for each ecological value and user interest chosen.
6   For critical time periods draw pressure-impact curves for each ecological value and user interest chosen.
7   MCA specialist loads the pressure-impact curves into an MCA tool to generate an optimal water level curve for each critical period and river reach or 'critical' river section. This is initially based on equal weighting of all functions and uses.
8   Expert panel weighs relative importance of impacts differently which may lead to identification of different optimal water level curves and environmental flow regimes, i.e. potential scenarios.
9   In this case, the hydropower producers calculated production level at different flow alternatives. Because of the national strategic importance of hydropower, the cost of foregone hydropower against optimal multiple use environmental flow alternatives was considered.

After trade-offs are completed, an optimum water-level curve is constructed and converted into water flow through use of hydrological models, to provide advice to hydropower companies on how to plan the dam release.

The advantages of this approach were reported to be that it:

- enables better documentation of expert judgement of environmental impacts of a range of different hydrological regimes;
- includes participation of all relevant stakeholders and defines the roles of the expert panel;
- is a quantitative approach which can work even in data scarce settings;
- is a computerised multi-criteria analytical tool that makes it easy to compare and co-weight the different river values/impacts;
- can be used both in setting water levels in reservoirs and as water flow in river reaches; and
- is flexible enough to be replicated and adapted to a variety of water management decisions, and updated at low cost as new information becomes available (Barton *et al.* 2010; Berge *et al.* 2010).

With the caveat that optimum water-level curve and critical periods may not be easy to decide for all river values due to lack of data, this approach could be used in a variety of situations. For example, options might be considered for supplying water flow to critical river reaches in critical time periods, such as waterholes for hippos in the dry season. With less available data, the results may be more ambiguous. While the option building based on pressure-impact curves could be used without MCA, MCA does allow for transparency about how criteria are weighted and decisions are made, important for good governance. In addition, trade-offs might be more difficult to make and 'more easily influenced by the strongest debater' (Berge *et al.* 2010: 116).

Another limitation of the MCA is that some techniques are extremely complex, must be undertaken by a specialist, and as such may detract from transparency or the decision-maker understanding the trade-offs. For the most part, though, even if the results of the MCA are not adopted, the method can still make a significant contribution to the decision-making process by better structuring the decision problem, making the process more transparent, and helping decision makers learn about inherent trade-offs.

For further information about MCA, Hajkowicz *et al.* (2000) provide a good explanation of the process, its strengths and weaknesses, and examples.

Impact assessment and MCA are among many tools that can assist in decision-making about water resource planning. An integral part of the many tools, is a final reminder about the important role of a consensus building approach in creating a total package that may involve trade-offs, but also results in mutual benefits.

A Sustainability Assessment Tool: The Global Reporting Initiative

The principles espoused earlier in this book reflect long-term triple bottom line economic, social and environmental concerns. Uptake of application and reporting using a range of sustainability assessment tools by corporate entities, particularly water utilities, has increased in both Australia and overseas over the past few years (Baldwin and Uhlmann 2010). A comparison of the comprehensiveness of six Australian and international business performance and sustainability indicator sets for urban water found that the United Nations Environment Program (UNEP) Global Reporting Initiative's (GRI) Sustainability Reporting Guidelines offered the best overall coverage of triple bottom line for any organisation to date (Uhlmann 2004). Of the various sustainability methodologies, the 2005 Public Agency Supplement to the GRI is the only method with the specific aim of achieving increased accountability in public sector performance. It is founded on the assumption that companies are accountable not only to shareholders but also to demonstrate to stakeholders progress in contributing to sustainability (GRI 2005). In reporting TBL performance, accountability is achieved through application of GRI principles about transparency, inclusiveness, auditability, completeness, accuracy, neutrality, and clarity, among others (GRI 2005: 22). Box 8.4 illustrates a possible application.

Box 8.4: GRI application to SEQ Water Strategy, 2009

The GRI was applied to the Southeast Queensland (SEQ) Water Strategy 2009, to compare options and to test workability of the methodology using publicly available data from reports and academic literature to investigate whether it could offer guidance in continuous improvement in delivering sustainable urban water supplies. In conjunction with the Public Agency Supplement, the GRI Guidelines enabled transparent assessment of both sustainability and decision-making. An ideal process for applying the GRI would be to have government collaborate with community in the assessment to enable access to greater data and 'provide a better informed and committed public, a publicly accountable assessment of urban water supply options, and a more sustainable outcome' (Baldwin and Uhlmann 2010: 200).

The application involved assessing beneficial and negative impacts of each of the eight water supply options from the draft SEQ Water Strategy 2009:

1 Demand management
2 Rain water tanks on individual houses or communal rainwater tanks
3 Stormwater harvesting
4 Recycled water: potable and non-potable
5 Desalination
6 Water grid
7 Harvesting the Mary River Reserve, via the Northern Pipeline Interconnector (previously represented by the disallowed Traveston Crossing Dam).

Eighteen core sustainability indicators were selected from the Public Agency Supplement, based on ready availability of data and those most relevant to the draft SEQ Water Strategy 2009. While there were other important indicators, relevant data applicable to each water supply option was not available, so they were excluded. The selected indicators were:

Social

1 Administrative efficiency and effectiveness of Government service (new social indicator, GRI 2005: 47), e.g. EIS quality
2 Reliability of water supply, potential for growth, and vulnerability to climate change (service quality standards, PR8)
3 Process for managing impact, e.g. community engagement (SO1)
4 Health and safety risks (PR1) of different water supply options

Economic

5 Cost of different water supply options (EC3)
6 Indirect economic impacts of water supply options (EC13)

Environmental

7 Total material use in water supply options (EN1)
8 Direct and indirect energy use by water supply options (EN3 & 4)
9 Total water use, ecosystems/habitats affected, recycling and reuse (EN5, 20–22)
10 Land used by water supply options (EN6)
11 Biodiversity impacts of water supply options, impermeable surface, protected areas/species, restoration (EN7, EN23–29)
12 GHG emissions of water supply options (EN8)
13 Waste generated by water supply options (EN11)

Public agency disclosures on policy and implementation

14 Relationship of agency within governmental structure (PA1)
15 Sustainable development (SD) definition used by government (PA2)
16 Aspects for which government has SD policies (PA3)
17 Short and long-term goals and targets for SD priorities, implementation, monitoring, and continuous improvement (PA4 & 6)
18 Process for setting goals and targets including stakeholder engagement (PA5 & 7).

Briefly, the assessment found that demand management is the least cost solution, saves energy and emissions. Domestic rainwater tanks are popular and the most favoured source of water for SEQ residents (Aquagen 2006; Nancarrow *et al.* 2007; White 2009). A $ AUD 1,000

Government household tank rebate is estimated to deliver water at a cost of $ AUD 0.54/kilolitres. The advantage of wide-scale decentralised rainwater harvesting, by these calculations, is from five to 18 times the cost benefit for the Water Grid. Even if the Government absorbed 100% of the cost of household rainwater harvesting installations, according to White (2009: 398) it would still be among the cheapest options for water supply. With improved demand management and localised supplies, an estimated 30% of potable water consumption within the commercial and industrial sectors could be saved at attractive payback periods (5–10 years), at less TBL cost than centralised energy-consumptive alternatives such as desalination plants (Werner and Hauber-Davidson 2008).

While indirect recycled water from treated sewage has been used successfully elsewhere for potable use, public acceptability is a major constraint affected by level of trust and risk management (Po *et al.*, 2003; Nancarrow *et al.*, 2007). Impediments to investment in recycled water also include development and energy costs, lack of financial incentives, risk of cross-contamination, complexity, and liability (Martin and Hill 2006).

Desalination plants have the second highest cost of development and operation but can yield a reliable long-term water supply with the flexibility to be decommissioned if not needed. However, a high proportion of growth in SEQ is inland, whereas plants are proposed to be located on the coast. More conclusive data comparing energy use and greenhouse gas emissions of recycled water treatment with desalination would add to options analysis.

A proposal to dam the Mary River would have used the 'Mary River Reserve'. Once the Traveston Crossing Dam was over-ruled by the Commonwealth government, the reserve was intended to be used as part of the northern interconnector of the Water Grid to pump Mary River water to Brisbane if needed. While the Grid had its benefits, the energy costs and greenhouse gas emissions of operation and moving water, water losses as well as potential risks from inter-basin transfers would need to be taken into account. The proposed Traveston Crossing Dam had only just been rejected, so it was included in this report as it had already been assessed. Of all the new water supply options, the assessment shows the disallowed Traveston Crossing Dam rated the poorest: it is most vulnerable to climate change due to variable streamflow, runoff and evaporation (from the shallow dam). At an early estimate of $ AUD 1.7 billion (QWI 2007), it was by far the most costly option to build even without considering indirect/embedded energy costs of dam materials, purchase cost and loss of prime agricultural land close to Brisbane, costs and energy related to a pipeline for transfer to Brisbane, impact on environmental flows and environmental impacts to threatened and endangered species, and displacement of the community and businesses in the Mary River Valley (indicators 3, 4, 5, 6, 7, 8, 10, 11, 12).

Assessment of governance indicators was also quite critical in terms of lack of transparency and inclusiveness in consultation and poorly presented and confusing information making comparability and auditability difficult (Baldwin and Uhlmann 2010: 199).

This study demonstrated the benefits of comparing water supply options even with limited data. When large amounts of public funds are being committed over the long term, policies mandating transparent consultative sustainability assessment of major infrastructure strategies are essential.

For greater detail on the assessment, refer to Baldwin and Uhlmann (2010). Much of the text in this box was taken directly from the article, with the authors' permission.

9 Monitoring and evaluation

Water resource planning should be seen as an ongoing cycle that adapts to improving knowledge and changing circumstances. In this chapter we consider performance indicators, monitoring programs and processes for periodic evaluation of water resource plans. Monitoring and evaluation are used for the ongoing improvement of the way a plan is implemented, and to inform periodic reviews of the plan itself. While we are discussing evaluation design as a late step in the planning process, in practice design of evaluation should commence at the step where objectives are being determined. It is presented in this chapter for convenience.

Purpose of monitoring and evaluation

No matter how much effort goes into planning, the reality is that no plan is likely to perfectly achieve its objectives because there will always be:

- actions that are not or cannot be implemented, or could be implemented more efficiently;
- uncertainty about the extent that actions will deliver the desired outputs and outputs will contribute to the objectives because of uncertainty about causal relationships or lack of sufficient data;
- changes in situational assumptions such as forecasts of climate, land use, and demands for water.

In addition, the appropriateness of objectives and community and government views on what is a reasonable compromise between competing objectives often change over time, as does knowledge about causal relationships and technology. For these reasons water resource planning should be seen as an ongoing cycle that adapts to improving knowledge and changing circumstances.

Recognising this, designing and funding a monitoring and evaluation regime is a core part of planning, but is commonly treated as a secondary consideration and not given the attention it is due. Consequently monitoring programmes are often poorly funded and poorly targeted, making meaningful evaluation difficult.

We argue that an up-front investment in design of monitoring and a commitment to meaningful periodic evaluation will yield substantial longer term benefits by reducing data collection that is of little benefit and improving the return on investment by improving efficiency and effectiveness of plan implementation and the plan itself over time. Monitoring and evaluation also provide accountability for invested resources and can provide an opportunity for further engagement and commitment from community.

Monitoring programmes assess the performance of the plan in relation to its outputs and objectives. They can guide adjustments to the way actions are implemented through periodic 'formative' evaluations. They can also inform less frequent 'summative' evaluations that consider the effectiveness of the plan as a whole in achieving its objectives and the appropriateness of those objectives, to determine whether a revision of the plan is needed, and guide the kinds of changes required.

In addition to learning from experience and adapting to change, evaluations provide a mechanism for accountability. They can show the extent that those responsible for implementing the plan have done so, and whether the investment required for making and implementing the plan has been sufficient and worthwhile. Often the effectiveness of a plan is reliant on the collaborative input of both government and stakeholder community, and the benefits can be greatly reduced if one of those parties fails to make their expected contribution.

Requirements for monitoring, evaluation, and review are incorporated into water resource planning policy and legislation around the world. In Australia, the NWI states that water resource plan performance will be monitored, factoring in knowledge improvements, and regular public reports will be provided, consistent with the nature and intensity of resource use (NWI s40). It also states that 'there should be a review process that allows for changes to be made in light of improved knowledge' (NWI schedule E, cl4). Consequently, all State government water resource planning legislation includes requirements for monitoring with a review of plans usually after a certain period of time – 5 or 10 years. In addition, such legislation usually prescribes a process for review, which may or may not be similar to the initial planning process that would have included stakeholder consultation.

Likewise UK guidance documents state that River Basin Management Plans (RBMPs) and their programmes of measures have to be reviewed after six years (DEFRA 2006: 56). It specifies that information used in the RBMP process should be updated to incorporate the most accurate information including changes in policy and legislation; environmental monitoring information; updated economic information; and experience gained from the previous planning cycles (DEFRA 2006: 56). Importantly, it specifies that the agency should draw upon stakeholders' experience in applying the previous RBMPs. One way of doing this is through a coordinated consultation process.

Similarly South Africa's *National Water Act* specifies that both the national water resource strategy and catchment management strategies must be reviewed at intervals of not more than five years (s5(4)(b); 8(3)(b)). Monitoring systems must provide for the collection of appropriate data and information to assess, among other matters, quantity, quality, use, and rehabilitation of water resources; compliance with resource quality objectives; and the health of aquatic ecosystems (NWA s.137(2)). The Act enables review of licences periodically as well.

Designing an evaluation

Monitoring and evaluation requires a commitment of time and resources by government agencies and the community. Being mindful of resource constraints, monitoring and evaluation should be designed so that it can be resourced and achieve as much value as possible. This means that it needs to be well targeted on matters that are of most importance, and be undertaken efficiently.

IUCN (2004) defines evaluation as

> a periodic assessment, as systematic and impartial as possible, of the relevance, effectiveness, efficiency, impact and sustainability of a policy, programme, project, Commission or organizational unit in the context of stated objectives. An evaluation may also include an assessment of unintended impacts.

It suggests five criteria to be used in IUCN funded evaluations:

Relevance – The extent to which the policy, programme, project or the organisational unit contributes to the strategic direction of IUCN and/or its members and partners. Is it appropriate in the context of its environment?

Effectiveness – The extent to which intended outputs (products, services, deliverables) are achieved. To what extent are these outputs used to bring about the desired outcomes?

Efficiency – The extent to which resources are used cost-effectively? Do the quality and quantity of results achieved justify the resources used? Are there more cost-effective methods of achieving the same result?

Impact – The changes in conditions of people and ecosystems that result from an intervention (i.e. policy, programme or project). What are the positive, negative, direct, indirect, intended or unintended effects?

Sustainability – The extent to which the enabling environment supports continuity of the policy, programme, project or work of the organisational unit. To what extent will the outcomes be maintained after development support is withdrawn?

The logic framework for water resource planning shown in Table 3.3 in chapter 3 (adapted from the World Bank (Team Technologies) Logic Framework Approach), suggests monitoring and evaluation at each level of the logic framework – actions, outputs and objectives. This is a different way of viewing evaluation to the IUCN 5 criteria structure. We suggest that, while the two approaches are different, they are compatible and can be merged into a structure that is helpful for water resource plan evaluation.

Considering this, in essence, a full 'summative' evaluation would consider:

1   Whether the plan objectives are appropriate (IUCN relevance, sustainability). This includes consideration of the relevance and consistency of the objectives, and the relative level of achievement of competing objectives, with government policies and to community aspirations. It may address whether the objectives as stated are clear in reflecting what was intended. It may review changes in government policies and community needs and aspirations since the plan was made to assess whether these require adjustment to the objectives.
2   How well the plan outputs are contributing to the objectives (IUCN effectiveness, impact). This can include confirming assumed causal relationships between outputs and objectives; and confirming whether assumed external actions and influences have transpired and whether unexpected external factors have impaired the achievement of the objectives.
3   How well the actions are delivering the outputs (IUCN efficiency, effectiveness). This can include confirming the causal relationships between actions and outputs; confirming assumptions about climate and water resource behaviour; and considering whether alternative actions might have been more efficient.
4   How well the actions are being implemented (IUCN efficiency). This includes examining the extent and efficiency of implementing the actions as set out in the plan.

While summative evaluations that consider the plan as a whole would consider all of these criteria, formative evaluations during plan implementation would focus on the third and fourth.

Within these four areas, evaluation questions can be determined that are appropriate to the particular circumstances of the plan from the plan's logic map. Given that a major source of information for evaluation will be the plan's performance indicators and corresponding monitoring program, these questions and the performance indicators should be developed at the time the plan is made, so the monitoring programme can be designed and implemented around them. Even so, the evaluation questions may be revised and updated at the time of either formative or summative evaluation to reflect newer priorities and concerns.

Given limitations on resources, a list of possible evaluation questions can be reduced to a shorter list of questions that are of most importance, considering criteria such as whether the question relates to:

• an area of high uncertainty (e.g. in the assumed causal relationships)

- an area of high public interest
- a need to demonstrate accountability
- an objective that is particularly sensitive to water management, as opposed to other factors
- an area where relevant information and expertise is or will be available.

For each evaluation question, performance indicators and information sources can be determined. This can be presented in a tabular form.

The examples in Table 9.1 are generic to illustrate the kinds of things that could be included. In practice, particular objectives, more specific questions, linked outputs, indicators and data sources would be included. Rather than covering all aspects of the plan, the selected evaluation questions could relate to objectives where there is greatest uncertainty and/or community concern, leaving out others where causal relationships are relatively certain. Agency staff should be familiar with the extent of significant policy changes or community concern about aspects of the plan.

Clearly the ability to develop a table such as this is heavily reliant on a well prepared plan logic table and a good knowledge of the nature and level of certainty of causal relationships that underpin it, and whether relevant monitoring information has been collected. Since many kinds of information (e.g. flows, economic performance, ecological condition) require the recording of information over extended periods of time, lack of this information will make responding to some evaluation questions difficult or impossible.

Running an evaluation

Ideally evaluations are conducted by persons who are independent of the government agencies responsible for implementing the plan, and are conducted in a manner that is objective, open and transparent to all stakeholders. This engenders confidence in the reliability of the conclusions drawn. Evaluations conducted by agency staff can still achieve much of this if the evaluation is done transparently and objectively, but they are commonly under implicit pressure to promote agency credibility and reduce perceptions of uncertainty (Allan 2008). We strongly advocate transparency and objectivity to build public confidence and trust and encourage adaptive learning, even when evaluations show that things did not work as well as was expected. In any case, views of and data from agency staff, need to be incorporated in the review, as staff should be familiar with the rationale (especially implicit) behind the initial plan and have good understanding of the challenges of implementation.

IUCN guidelines (IUCN (2005: 23) recommend that a good evaluator or team of evaluators will have:

- expertise in methodology of evaluation
- thematic expertise
- credibility with stakeholders

Table 9.1 Evaluation questions and indicators

Area	Evaluation questions (generic examples)	Performance indicators (generic examples)	Information sources (generic examples)
Whether the plan objectives are appropriate	Are plan objectives aligned to government policies? Do plan objectives reflect current community priorities and aspirations? Is the level of risk associated with any objectives leading to impacts that are no longer acceptable? Are there important benefits and beneficiaries not addressed in objectives?	Cases of inconsistency with government policies or initiatives. Cases of relevant policies not addressed in the plan. Difference between current and forecast consumptive demands for water and those that were assumed previously. Stakeholder satisfaction with objectives. Stakeholder concerns about impacts.	Analysis of current government policies. Advice from relevant agencies. Assessments and projections of consumptive water demands. Community surveys and submissions.
How well the plan outputs are contributing to the objectives	Are objectives being achieved? Are plan outputs contributing to objectives as per assumed causal relationships? Are key externality assumptions close to what was expected?	Various direct or surrogate indicators of achievement of objectives (economic, environmental, social benefits). Extent that causal relationships are confirmed by studies. Comparative estimates of achievement of objectives with and without plan outputs. Comparison of actual to predicted external factors.	Monitoring of economic, environmental and social condition. Research into causal relationships. Models that estimate change in achievement of objectives with and without the plan. Analysis of externalities. Community surveys and submissions.
How well the actions are delivering the outputs	Are outputs being achieved? Are plan actions contributing to outputs as per assumed causal relationships? Are assumptions about climate, water resource and water user behaviour reasonable? Are there alternative or modified actions that could deliver the outputs more efficiently?	Various means for directly measuring achievement of outputs. Extent that causal relationships between actions and outputs are confirmed by studies. Comparative estimates of achievement of outputs with and without plan actions. Comparison of actual to predicted climate and water resource behaviour, and changes in forecasts for future climate. Identified alternative or modified actions that are more efficient.	Monitoring of water regime (rainfall, flows, levels, storage, etc.) and water extraction (metering). Updated water resource modelling and climate forecasting. Research into causal relationships. Models that estimate change in achievement of outputs with and without the plan. Agency staff and community surveys.
How well the actions are being implemented	Are actions being implemented efficiently?	Extent to which actions are implemented. Identified options to improve cost-effectiveness of implementation.	Audits of implementation of actions. Agency staff and community surveys.

- impartiality – in particular no conflict of interest with any interested parties
- communication skills – to communicate evaluation results so they can be understood by all interested parties
- interpersonal skills so as to be able to interact with parties in a sensitive and effective way
- availability to conduct the evaluation in the time frame.

As noted above, information for evaluations typically comes from a combination of monitoring programs that collect data over an extended period, scientific studies, modelling, engagement with agency staff and input from community stakeholders. Thus an evaluator will require access to this information and willing collaboration from agencies who hold the information. They will require the resources and time to collate and analyse information and to engage with agency staff and the community to obtain needed input.

The evaluation will seek to provide answers to the selected evaluation questions and justify them with evidence. It will identify areas where the plan has been successful and areas where improvement or change should be considered. A legitimate conclusion within an evaluation report is that there is insufficient information to address a particular evaluation question and recommendations to address that for future evaluations.

Using evaluation results

Evaluations can positively improve water resource management, provided they are used. This requires institutional and governance arrangements that support an adaptive management paradigm, with adjustment of implementation reflecting ongoing regular formative evaluations (typically every one or two years, or event based), and whole-of-plan periodic summative evaluations (typically at 5 to 15 year intervals). Where there is substantial government or community cost involved in plan implementation and considerable uncertainty about whether the actions will deliver the expected outputs and objectives, the cost of monitoring and evaluation and the need for a culture of adaptation can be readily justified. To mitigate against the tendency of governments to not admit uncertainty and failure, it is essential for assumptions and level of uncertainty or unreliability of data, and the rationale for monitoring, to be clearly stated during initial plan development. Long-term community trust and confidence requires openness and honesty, including a willingness to adapt based on new evidence.

Designing performance indicators and monitoring programmes

As discussed in chapter 6, design of performance indicators commences at an early stage of the water resource planning process. We discuss it more fully in this chapter because of its strong linkage to the design of monitoring and evaluation.

Performance indicators define how success in achieving objectives and outputs will be evaluated, and enable this to be quantified in targets. The World Bank (Team Technologies 2005: 37 onwards) describes performance measurement as the process of 'identifying the features that define the actual quantitative, qualitative, timing, cost and place parameters' for the plan. Performance indicators 'define and measure objectives'. They are a 'description of results, not the conditions necessary to achieve them'. The Bank recommends that:

- Only the number required to clarify what is to be accomplished should be used. In general the fewer the better.
- Industry standard indicators should be used where available and practical.
- Indicators should be measurable in terms of quantity, quality and time. Sometimes location and cost can also be added. A four step process is suggested, illustrated by this example:

  1 Basic Indicator: Rice yields of small farmers increased.

  2 Add Quantity (how much): Rice yields of small farmers increased by x bushels (or from x to y).

  3 Add Quality (what kind of change): Rice yields (of same quality as 199x harvest) of small farmers (owning 3 ha or less) increased by x bushels (or from x to y).

  4 Add Time (by when): Rice yields (of same quality as 199x harvest) of small farmers (owning 3 ha or less) increased by x bushels (or from x to y) per annum starting in 199x harvest.

- Indicators must be practical in terms of ease of use, affordability and attribution. Is there a methodology and expertise to collect the necessary monitoring data? Is it affordable? Do the indicators really explain what is intended to be achieved?
- Where achievement of outputs and objectives can only be assessed on a long-term basis, leading and process indicators may also be important. Rather the measuring the result, leading and process indicators signal whether there is progress being made that should contribute to the result in order to give confidence (or raise the alarm) during implementation.
- Proxy indicators may be used where more direct indicators are too difficult or expensive. For example measuring rural income levels may be difficult. Possible proxy indicators might be suggested by stakeholders, for example the number of televisions owned.

GWP (2006) notes some common pitfalls in developing indicators and monitoring programs:

- Not having a system: defining a loose collection of disparate indicators with little or no relationship to each other, instead of a system in which indicators relate to each other and to the [plan objectives and outputs] in a meaningful way.
- Bad fit between targets and indicators: Defining indicators with a weak relationship to the targets set for [outputs and objectives]. In most cases, the problem is with the indicator; in others, the root of the problem is a poorly formulated target.
- Building a system based on poor baseline data and/or unreliable indicators: indicators need to provide a consistent measure of progress. This means that the starting point (the baseline data) are accurate and that the indicator provides an objectively verifiable result, i.e. two people applying the same indicator should get the same result.
- Not taking into account that impacts may differ according to location and to the gender and socio-economic status of intended beneficiaries.

In many cases, the greatest liability in developing a monitoring programme is the deficiency in existing long-term consistent and reliable data about water use/extraction of water, water and ecosystem resources, and relevant socio-economic variables. So in developing a water resource plan, commitment might need to be sought for basic equipment and training, for example, for metering on all pumps extracting more than a certain amount, and reliable, well instrumented, river gauges at critical locations.

To ensure that time and resources are not wasted on irrelevant monitoring and that monitoring is able to inform effective evaluation, the selection of performance indicators and monitoring programmes and tools must be closely aligned to the logic model of the plan and to those particular aspects that are selected as a focus for evaluation.

Of the four areas of evaluation questions in Table 9.1, indicators and monitoring programs for the second and third areas are discussed as these are the ones where advanced planning and long-term investment is most commonly needed to deliver meaningful performance information.

Plant *et al.* (2012) propose that selection of appropriate performance indicators and monitoring can be informed by using information from an ecosystem services assessment done during situational analysis along the lines of that shown in Figure 5.2. Table 9.2 illustrates their approach. It assumes that objectives and outputs have been expressed using ecosystem services assessment information, as was discussed in Chapter 6.

Plant *et al.* (2012: 56) comment that the cost and effectiveness of possible monitoring programmes can be a deciding factor in selecting performance indicators, and refer to this example.

Table 9.2 Deriving performance indicators and monitoring from ecosystem services assessment

	Objective			Output
	Beneficiaries	*Benefits*	*Services*	*Supporting water regime*
Objective / output	Support commercial and recreational fishers, Indigenous communities, and businesses associated with recreation and tourism...	...by maintaining and improving fish catches...	...through maintaining habitat and flows for fish breeding and growth in the areas shown on map x...	...by protecting low flows and freshes to support fish breeding and growth.
Performance indicators	Satisfaction of fisherman with fishing opportunities and catch exceeds pre-plan levels.	Number/weight of fish caught by recreational fishers equals or exceeds pre-plan levels.	Populations of fish species x,y,z in areas on map j increase by 10% and fish habitat is improved and increased in areas on map h over the next six years.	Low flows below the 80th percentile, and bank full or higher natural flood flows at least once every two years are protected in reaches shown in map j.
Monitoring programme	Annual survey conducted by fishing club. Question added to surveys conducted by tourism and recreation organisations.	Annual survey conducted by fishing club. Question added to surveys conducted by tourism and recreation organisations.	Annual fish habitat surveys using method x. Monthly fish population monitoring at selected indicator sites.	Extraction of water during times when estimated natural flows would be less than the 80th percentile. Frequency and duration of bank full or higher flow events at indicator sites.

(Adapted from Plant *et al.* 2012: 57. Used with permission.)

In [the example shown in Table 9.2], the identified option for benefit monitoring (annual satisfaction survey by fishing club) is risky in the sense that it may not deliver results that are statistically sound. On the other hand, the water regime monitoring proposed could be done for a variety of other purposes, so is both cost-effective and highly objective. Fish habitat monitoring (the service) might be cost-effective especially if done for other purposes, but even if not, it may be worth funding, if fishing is important to the area or recreational fishers are a sensitive stakeholder group.

In addition, they state that it is important to understand the role of extraneous factors in achieving outcomes. Whether monitoring of these factors is necessary should be explicit and depends on sensitivity to these other factors.

In the example shown, if the performance indicator of 'satisfaction of fishermen with fishing opportunities and catch' (benefit) is chosen, then consideration should be given to what other factors could impact on satisfaction, and whether these can be separated out in evaluation. This may lead to additions or modifications to the monitoring programmes. In this case, the survey of fishermen could request information on factors impacting satisfaction, in addition to fish populations (e.g. was dissatisfaction due to cuts in access to fishing areas, drought, cattle trampling vegetation, etc.?). Similarly the performance indicator relating to fish habitat monitoring might be supplemented by water quality monitoring. (Plant *et al.*: 57)

Another consideration raised by Plant *et al.* (2012) is time lags. They refer to the example to illustrate that monitoring and evaluation needs to be imbedded as early as possible in the 'supply' chain in order to be responsive.

For example, surveying beneficiaries about the delivery of a benefit they receive can sometimes be too late in the supply chain – when a fisherman notices a consistent decline in catch after a number of seasons, it may be the manifestation of flow conditions several years previously that prevented breeding. In other cases time lags are shorter – when a grazier notices his stock's condition declining due to the poor condition of floodplain fodder, it may be due to poor floodplain inundation last season. Monitoring early in the supply chain will give an earlier signal that a benefit might change, and again a service-based indicator may be responsive and still retain meaning for beneficiaries, especially when communicated as part of a linked chain. (Plant *et al.*: 58)

Monitoring of the water regime and water extraction is a general core requirement. This includes river level and flow gauges, storage gauges, rainfall gauges, and groundwater level monitoring bores. The required frequency of measurement to support modelling and assessment is usually continuous or daily for surface water. River flow measurement usually requires the establishment of a relationship between water level and river flow at a point (a flow rating curve). As river beds are often unstable this relationship can be unstable also, requiring an ongoing program of field flow measurement to check and recalibrate the rating curve. It can be worthwhile to invest in a structure to stabilise and improve the sensitivity of the rating curve at low flows, as shown in Figure 9.1 (see colour plates).

Groundwater is commonly monitored on a monthly or quarterly basis, but may need higher frequencies in some cases. Monitoring of water extraction can be via meters on pumps, bores or channels, or where this is not feasible through various estimation techniques such as those using areas of crops irrigated, pump hours, etc. This information underpins understanding how the water resource behaves, climatic analysis, water user demand analysis, and the development of models that are able to support decision-making.

Performance indicators for objectives often include such things as economic outputs and ecosystem condition, with associated monitoring programmes drawing on the broad state of the environment, economy and society information that is published and referred to by governments for multiple purposes. This is cost-effective, but as shown in the discussion above, their value for evaluating water resource plans can be limited because of the difficulty in separating out the effect of externalities from the effects of water management.

For example, economic outputs from irrigated agriculture are often greatly impacted by commodity markets, pests and diseases and farming practices. For water resource planning there needs to be a means to separate the effects of water resource management from other factors in analysing fluctuations in economic output levels. One approach to doing this is to use models that are developed to simulate water availability and production over a period of time. Such models are used in assessing possible management options using forecast future climate scenarios (see Chapter 8). They can also be used with current and historic climate data to estimate the effect of the implemented management actions as compared to what would have occurred without them. As with all models, the weakness is the extent that the models are able to accurately simulate complex relationships, which varies substantially for different circumstances.

There are a great many different schemes for the monitoring of the condition of river and wetland ecosystems. A well-known example in Australia is the annual monitoring of the ecosystem health of Southeast Queensland catchments, undertaken by the not-for-profit Healthy Waterways organisation, as shown in Box 9.1.

Box 9.1: South East Queensland Ecosystem Health Monitoring Program, Australia

This is a well-established and ongoing monitoring programme for water resource condition operating in south east Queensland, 'to assess the effectiveness of management and planning activities aimed at improving SEQ's waterways' (SEQHWP 2010). The programme reports on the health of freshwater ecosystems and estuary/marine ecosystems. The freshwater health rating is based on combining five indicators:

- Physical and chemical indicator (combining pH, conductivity, temperature, dissolved oxygen);
- Nutrient cycling indicator (combining nitrogen stable isotope signature, algal bioassay of nutrients);
- Ecosystem processes indicator (combining algal growth, carbon stable isotope signature, benthic metabolism);
- Aquatic macro-invertebrates indicator (combining number of taxa, PET richness, SIGNAL score);
- Fish indicator (combining percentage of native species expected, ratio of observed to expected native species, proportion alien fish).

The results are aggregated at a subcatchment level. A graphical presentation of the results is shown in Figure 9.1 (see colour plates).

The visual portrayal of the Report Card data and accompanying detailed water quality monitoring assists in prioritising efforts by municipalities and land management agencies of where to focus resources such as upgrading sewage treatment facilities and riparian rehabilitation. Thus it is used on an ongoing basis to refine management, not just at plan review periods. It is also a key mechanism for ongoing community awareness and engagement, fostered more recently by the ability to download conditions as an iPhone app. For the latest status, we refer readers to the Ecosystem Health Monitoring Program on www.healthywaterways.org.

Like economic monitoring, ecosystem condition monitoring programmes may not be helpful for evaluating water resource plans if they are unable to differentiate effects of factors outside of the scope of the water resource plan from effects of the water resource plan. One way of addressing this is to use thorough scientific studies in representative locations to establish and confirm causal relationships between water regime characteristics (outputs of a water resource plan) and ecosystem condition (objectives) with other factors kept constant, then rely solely on monitoring outputs that are known to be conducive to the desired ecosystem condition. For example, consider the chain of action/output/objective shown in Figure 9.3.

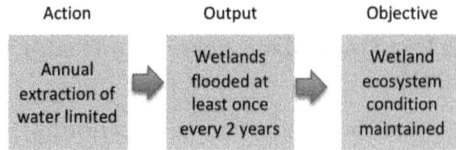

Action	Output	Objective
Annual extraction of water limited	Wetlands flooded at least once every 2 years	Wetland ecosystem condition maintained

Figure 9.3   Example chain of action/output/objective

Direct measurement of wetland ecosystem condition may be possible, but might not be useful as far as evaluating the effectiveness of the water resource plan is concerned because it can be impacted by matters that may be outside the scope of the water resource plan, e.g. grazing by stock, encroachment of introduced species, etc. This is one of the reasons for good linkages and integration between agencies responsible for water resource planning and those dealing with broader catchment matters. Additional studies can confirm whether a targeted flooding frequency will deliver the expected ecosystem response such as maintenance of fish or vegetation, all other matters being addressed, then use the planning provisions to monitor only flood frequency. This, for example, is applicable to the Murray-Darling Basin in Australia where the Commonwealth Environmental Water Holder manipulates the amount of water delivered to certain high priority wetlands, and needs to know whether the costly process (due to buy-backs and productive opportunity lost) is effective.

Finally, selection of performance indicators will require hard decisions to be made about what will deliver the most relevant information for answering evaluation questions, and are at the same time able to be funded and implemented given available skills and resources. It is unlikely that all desired indicators will be able to be monitored. Careful consideration and analysis as outlined above can lead to selection of indicators and monitoring that will deliver the best value for money.

Around the world, there is seldom sufficient funding to undertake all the monitoring required. Community-based organisations can provide valuable input, with support of training programmes to ensure consistency and reliability of data over time and among data gatherers. Community monitoring builds awareness of resource condition and effectiveness of management, and provides a vehicle for stewardship and adaptive management of the resource. Some examples follow to illustrate the different ways of implementing and using such groups for monitoring.

Box 9.2: Waterwatch Australia

Waterwatch Australia is an overarching national volunteer water quality monitoring and education organisation with separate programmes in each state. Community groups and schools in collaboration with councils, government agencies and regional NRM bodies develop strategies to deal with water quality issues in their catchments. The Commonwealth Government has provided funding for Waterwatch coordinators and education projects.

In NSW, twice a year hundreds of schools, community groups and individuals sample the range of macro-invertebrates (water bugs) present in local waterways. In 1997 over 20,000 people searched high and low for insects, crustaceans, molluscs, and worms. The information collected during the NSW 'Water Bug Survey' is collated and provides a snapshot of the health of waterways in NSW.

In 2007, Queensland government agencies produced a manual for the use of community groups like Waterwatch to provide a strategic approach to waterway monitoring and enhance data confidence. It includes advice on designing a study, methods for monitoring water quality, biological and stream condition and habitat, as well as data management and reporting. It includes record sheets with visual prompts and indicators (DNRW 2007).

Box 9.3: Community monitoring of impact of Pak Mun Dam on downstream subsistence

The Pak Mun Dam, situated on the Mun River close to its confluence with the Mekong River, was completed in 1994 by the Electricity Generating Authority of Thailand to generate hydroelectricity. Community concerns about the impact of the dam on the downstream subsistence fishery, led to a trial opening of eight sluice gates from April 2001 to June 2002 to allow studies on fisheries, social impacts and impact of the dam on Thailand's electricity supply. Community-based monitoring contributed to the assessment.

Villager-led Thai Baan research, 'grassroots people's research', was conducted with the help of the Southeast Asia Rivers Network. It documented fish and flora species and fishing gear and demonstrated the communal values and culture in relation to the river. The economy of villagers below the dam had depended exclusively on fishing because there is little land suitable for agriculture. It is thus the most important resource base for the majority of people in the region – landless villagers and those who possess small amounts of land.

The study used inclusive consensus-based approaches to research, local researchers, and formed an expert panel of locals (e.g. fishermen). The Thai Baan researchers included 200 villagers from 65 communities nominated to collect data by the local communities for their expertise in different fields. They received no compensation and were assisted by NGO advisors. The research was empirically based on field observations and the participation of local people, with group discussions to ensure the accuracy of the analysis. The research initially focused on the impact of the opening of the dam gates on fisheries, however given links to a broader context, the researchers extended the research to also address:

- river ecosystems
- plant and vegetation
- fishing gear
- dry season river bank vegetable gardens and
- social, economic and cultural issues in the local context.

The methodologies applied to study each issue involved observation, in situ recording of field observation, validation of data by local experts, data classification and analysis. This included interviewing and documenting villagers who had left for employment in other regions of Thailand but who returned to the community after the opening of the sluice gates.

It confirmed the benefits of the 2001–2002 opening of the gates in terms of fisheries, ecology, vegetable production, tourism, and cultural events (Southeast Asia Rivers Network 2004). Almost every household in the study area had reduced food expenses as a result of fishing.

The community-based Thai Baan research illustrates the immense potential for harnessing local knowledge and mechanisms for monitoring that can provide input into dam operating systems. It demonstrates available skills and resources within the country. The studies of the fisheries, social impacts and impact on electricity supply confirmed that crises and disputes can be turned into opportunities if the responsibility of managing resources is shared among people (Assembly of the Poor – Pak Mun 2002).

As can be seen in Table 9.1, research into causal relationships may often be required to evaluate a water plan. This is ideally focused on areas of greatest uncertainty and risk that are identified during the plan making process. For example, if there is uncertainty about the water regime requirements for maintaining water quality, or for the preservation of a valued fish species, then this might be the target of research that is done as part of plan implementation.

Lastly, monitoring programmes often have an operational function during plan implementation. For example, river levels and flows are commonly used for operation of dams and for regulating access to water by consumptive water users, to provide for protection of environmental water needs and for sharing of water between water users. There can also be operational triggers set in the plan to allow for uncertainty in predictions of climate, water resource behaviour or ecological response.

For efficiency, design of monitoring programmes seeks to achieve both operational and longer-term performance evaluation purposes in the most integrated way possible.

Completing the logic model

Finalisation of performance indicators, monitoring programmes and evaluation approach allows the logic model set out in Table 3.3 (Chapter 3) to be completed. This sets out transparently and logically what the water resource plan intends to achieve, how it intends to do so, key assumptions about matters outside the scope of the plan, and how success will be measured and evaluated.

The entire logic framework aims to identify priorities and strategically place resources to get the most benefit, within the limit of resource constraints. Monitoring and evaluation complete this providing for ongoing adaptive improvement.

10 Conclusion

Based on our combined experience working in government, consulting and research, it became clear that there was need for a book which could provide practical guidance on water resource planning for those conducting a planning process, those engaged in the process as well as for policy analysts who are establishing an institutional framework for water resource planning. By taking readers through every step of the process, we intended to not only provide insights on a variety of ways to achieve good practice, but also to show how core principles can be embedded in the process.

Throughout the book we draw repeatedly on different aspects of case studies from Australia (especially the well-known Murray-Darling Basin), the European Union, and South Africa, each of which has used a different model for reforming the water allocation process. Each of these are built on principles designed to achieve equitable distribution of the benefits of water for humans and the environment, procedural fairness in community input to decision-making, and long-term sustainable use. Each case study agency also has a regular system of review to assess progress on individual catchments as well as across the country or countries. This enables adaptive adjustment and improvement of institutional arrangements and management approaches at a catchment scale. We also refer to other examples from around the world to illustrate possible approaches and actions, to demonstrate that each process needs to be, and can be, tailored to its own purpose and context, and that there is often not a 'one-only' solution to a problem. We aim to provide sufficient evidence to enable water planners and policy advisors to state a claim for reform and good practice. Most importantly we encourage planners not to be afraid to try a different approach from the status quo, provided it is launched within a spirit of community engagement, monitoring of effectiveness, reflection and adaptive management.

This final chapter synthesises characteristics of global principles for best practice water resource planning. It summarises how these are addressed through a logical framework and systematic approach to provide guidance along the often bumpy road, strewn with potholes and obstacles, to achieve sustainable outcomes. It identifies features that should be targeted for improvement and for further research in this area.

Four basic challenges to achieving the goals of integrated water resource management were identified in Chapter one: 1) Better use of existing resources; 2) Ecosystem and environmental quality; 3) Uncertainty about the future and implications for water security; 4) Conflict due to inequitable distribution of costs and benefits about water allocation. We argue that these challenges can be addressed in a logically set out water resource planning process, in a context of appropriate enabling governance and policy, and we describe guidelines for doing so. We suggest there are advantages to applying these guidelines to transboundary arrangements as well as within state cases, to provide a common focus for collaborating parties. We are concerned that while most agreements describe collaboration processes, many focus on water quality and do not include clauses or much detail about water allocation, yet perceived inequity of distribution is often a basis for conflict between upstream and downstream users.

A prerequisite and overarching first guideline for good water resource planning is legislation and policy that establishes *Integrated and participatory governance for water resource planning and management*. These institutional arrangements should ensure an appropriate scale for planning, equitable sharing of benefits and responsibilities, and multi-level commitment to cooperation among agencies, community engagement, transparency, and accountability.

Another guideline is having sufficient *knowledge* of the resource and its use on which to base decision-making. Adequate data on inputs, outputs, behaviour of the water resource and estimated futures based on demands and uncertainties are the foundation that enables understanding of the social, cultural and economic relationship with the resource. There needs to be understanding of the benefits associated with water supply, non-consumption, and Indigenous and disadvantaged people as well as the relationship between these benefits and the characteristics of the water resource. In addition to scientific assessments, partnerships with the community can assist in filling information gaps based on experiential knowledge of hydrologic and ecosystem behaviour, as well as socio-economic information. This is normally compiled during the *situational analysis* stage early in a planning process, with information added as identified knowledge gaps are addressed, or supplemented later when monitoring progresses.

Getting agreement among parties about desired *outcomes and objectives* can provide common goals that can be used to keep discussions focused, for example particular local economic, social and environmental benefits. While these are frequently established early in the process, often based on legislation and government policy, it is important that there is transparency around whether they can be achieved, or rather, need to be adjusted to be more realistic. The logical relationships between the broader benefits (and beneficiaries) set out in objectives and water resource characteristics should be clearly described so that the expected results of water resource management can be clearly understood by planners and the community.

This will likely happen in tandem with developing and assessing a wide range of possible actions to achieve the objectives. These can include storage release operating rules, regulation of extraction through licensing and metering, self-governance, and economic measures. The benefits of engaging the community at this stage include identification of a range of innovative and locally appropriate actions and rigorous testing of the thoroughness of assessments.

Finally the importance of monitoring and review of the effectiveness of management actions in individual plans in achieving objectives cannot be overstated. Well-structured monitoring and evaluation can be cost effective in adapting to changing circumstances and knowledge and in improving the return on investment by governments and the community into water resource management. Including community in progressive implementation of the plan as well as monitoring and review processes can inform corrective action as needed and refine adaptive management and stewardship of the resource.

Features for improvement and further research

In a world of ever-increasing demands for limited government funding, investment in water resource planning is constrained and in many cases being reduced. Most vulnerable to the red pen is funding for data collection, research and monitoring, and investment of time and resources in community engagement. Yet the water industry knows that defensible data and engagement and inclusion of local knowledge contributes to assessing trade-offs and making better decisions. Such decisions can be more easily implemented and gain more effective acceptance and compliance if the community is involved. A system that avoids engagement is ripe for less justifiable or accountable decisions, aligned with those with greatest influence. On the other hand, clever inclusion of the community can cost-effectively assist in data-poor areas, derive innovative solutions, and reduce later costs of dissatisfaction and conflict. To be consistent with Agenda 21 and the principle of subsidiarity, developing capacity at the lowest appropriate level is a key ingredient to success in achieving sustainable development.

Given the limitations in funding, it is important that what is done is efficient and effective to ensure the maximum return on investment. Our observations and assessment of water planning in Australia and other areas suggests that there are many opportunities for improvement that are not dependent on increased funding. These include:

- having in place guiding principles that ensure that broader community objectives are properly considered;
- identifying all beneficiaries and benefits (including non-consumptive), that could be affected by the plan, and clearly stating plan objectives that encompass these benefits, so they can be considered from the start rather than being an afterthought that causes difficulties late in the planning process or during plan implementation;

- design of community engagement methods to fit the purpose and the stakeholders' requirements, rather than applying an ill-fitting standard approach;
- focusing research around benefits that are most at risk and least understood, rather than having scatter gun research programs that may not focus on where the need is greatest;
- identification of multiple future scenarios of climate and development, to reduce risk and cost of failure during plan implementation because no thought was given to coping with things such as worse than expected droughts;
- defining the plan logic early, so that actions can be focused on key areas and all understand the relationship between plan actions and the benefits and beneficiaries;
- having an open process for identifying and assessing management options that can lead to selection of options that are innovative, locally appropriate and cost effective, rather than simply applying standard actions for all cases;
- ensuring there is procedural fairness in the process from the start, thereby reducing time and resources required to address actual or perceived inequities late in planning and during implementation;
- planning for adaptive management in ongoing plan implementation, so that actions can be adjusted 'on the fly' to address problems in implementation, improved knowledge and local variations;
- establishing cycles of longer term evaluation and reviews of plans, so that plans stay relevant and reflect the latest knowledge and experience;
- establishing monitoring programmes that strategically relate to relevant and practical performance indicators, rather than poorly targeted monitoring programmes that prove to be of little value for evaluation.

We believe that governments and other bodies that are conducting or supporting water resource planning would find it worthwhile to develop practice guidelines that encompass the above matters in a way that is relevant to their local water resource planning framework and needs. This will ensure that planners have practical guidance in methods and approaches for plan development, and reinforce a commitment to the process from higher organisational levels.

Across the globe, water planners are challenged with how to incorporate the needs and values of under-represented groups and Indigenous peoples in water planning. In Australia, a clear intent of the National Water Initiative and the National Water Commission is to engage Indigenous people in water resource planning to address their cultural and economic needs. While there has been an improvement in engagement, delivering their needs has proven a challenge given the boundaries of legislative processes, the history of land and water development in Australia, and marginalisation of Indigenous people. This does reinforce the need for collaborative processes that make links outside of the regulatory system where at least some of their needs might be addressed. In Australia, as in other countries, this includes multi-level governance arrangements across government agencies, and non-governmental catchment or river basin level agencies.

We hope that, should we (or others) prepare an updated version of this book at some time in the future, many more examples of good practice will be included. Although there has been major progress in implementing water reform around the world, we limited our examples to Australia and a few other regions, and acknowledge that more examples exist and will be reported over time. We suggest that assessment according to these guidelines will help identify good practices.

We hope that the principles, guidelines, and logic framework for water resource planning, supported by examples, will lead to improved outcomes for the resource, benefits for humankind, and less conflict in getting there.

Bibliography

Abseno M, 2009, 'The concepts of equitable utilisation, no significant harm and benefit sharing under the Nile River Basin Cooperative Framework Agreement: some highlights on theory and practice', *Journal of Water Law*, Special issue: Promoting water for all, 20 (2/3): 86–95.

ADB (Asian Development Bank), 2006, *Rehabilitation and Management of Tanks in India: A Study of Select States*, ADB, Manila, Philippines.

Allan C, 2008, 'Can adaptive management help us embrace the Murray-Darling Basin's wicked problems?', in C Pahl-Wostl, P Kabat, and J Möltgen, (eds.) *Adaptive and Integrated Water Management: Coping with Complexity and Uncertainty*, Berlin: Springer.

Amengual M, 2006, *Incorporating Local Knowledge into Joint Fact Finding*, Massachusetts Institute of Technology, viewed 30 October 2006, <web.mit.edu/dusp/epp/music/>.

Amer M, Daim T, and Jetter A, 2013, 'A review of scenario planning', *Futures*, 46: 23–40.

Aquagen, 2006, *Future water sources for the sunshine coast: community consultation*, June 2006.

Arnstein, S, 1969, 'A ladder of citizen participation', *Journal of the American Institute of Planners*, 35(4): 216–224.

Arthington A, Brizga S and Kennard M, 1998, *Comparative Evaluation of Environmental Flow Assessment Techniques: Best Practice Framework*, Canberra: Land and Water Resources Research and Development Corporation.

Arthington A, Naiman RJ, McClain ME and Nilsson C, 2010, 'Preserving the biodiversity and ecological services of rivers: New challenges and research opportunities', *Freshwater Biology*, 55: 1–16.

Arumugam B, Mohan S and Ramaprasad R, 1997, 'Sustainable development and management of tank irrigation systems in south India', *Water International*, 22(2): 90–97.

Aslin H and Brown V, 2004, *Towards Whole of Community Engagement: A Practical Toolkit*, Canberra: Murray-Darling Basin Commission, viewed 12 November 2013, http://www.mdba.gov.au/sites/default/files/archived/mdbc-S-E-reports/1831_towards_whole_of_community_engagement_toolkit.pdf.

Assembly of the Poor-Pak Mun, 2002, *Progress Report of Thai Baan Research: Findings of Community-based Research on the opening Pak Mun Dam Gates in Thailand, viewed 3 August 2006, http://www.internationalrivers.org/campaigns/pak-mun-dam–0.*

Åström J, Granberg M and Khakee A, 2011, '"Apple pie-spinach" metaphor: Shall e-democracy make participatory planning more wholesome?', *Planning Practice & Research*, 26(5): 571–586.

AZDWR, 2010, *Draft Water Demand and Supply Assessment for Phoenix Active Management Area*, Arizona Department of Water Resources, November 2010.

AZDWR, 2011, *Phoenix Active Management Area Water Demand and Supply Assessment 1985 to 2025 – Summary*, Arizona Department of Water Resources, December 2011.

Baker M, Coaffee J and Sherriff G, 2007, 'Achieving successful participation in the new UK spatial planning system', *Planning Practice & Research*, 22(1): 79–93.

Baldwin C and Sjah T, 2012, 'Water governance in East Lombok: the feasibility of co-management to resolve competing uses', *Tapping the Turn: Water's Social Dimensions*, 15–16 November 2012, Canberra.

Baldwin C and Twyford V, 2007, 'The Challenge of Enhancing Public Participation on Dams and Development: A Case for Evaluation', *International Journal of Public Participation*, 1(2): 17–23.

Baldwin C and Uhlmann V, 2010, 'Accountability in Planning for Sustainable Water Supplies in South East Queensland', *Australian Planner*, 47 (3): 191–202.

Baldwin C, Hamstead M and Uhlmann V, 2008, *Interjurisdictional analysis of self-management governance arrangements for water resource management in Western Australia*, Report to Department of Water, Perth, Western Australia, 24 October 2008.

Baldwin C, O'Keefe V and Hamstead M, 2009, 'Reclaiming the balance: social and economic assessment – lessons learned after 10 Years of water reforms in Australia', *Australasian Journal of Environmental Management*, 16(2): 70–83.

Baldwin C, Tan P-L, White I, Hoverman S, and Burry K, 2012, 'How scientific knowledge informs community understanding of groundwater', *Journal of Hydrology*, 474: 74–83.

Barber K and Rumley H, 2003, *Gunanurang:(Kununurra) big river, Aboriginal cultural values of the Ord River and wetlands*, Report prepared for the Water and Rivers Commission, Perth.

Barma D and Varley I, 2012, *Hydrological modelling practices for estimating low flows – guidelines*, National Water Commission, Canberra.

Barnett B, Townley LR, Post V, Evans RE, Hunt RJ, Peeters L, Richardson S, Werner AD, Knapton A and Boronkay A, 2012, *Australian groundwater modelling guidelines*, Waterlines report, National Water Commission, Canberra.

Barrett C, 2009, *Lower Gwydir Groundwater Source: Groundwater Management Area 004 Groundwater Status Report 2008*, NSW Department of Water and Energy, Sydney.

Barton D, Berge D and Janssen R, 2010, 'Pressure-impact multi-criteria environmental flow analysis: Application in the Oyeren delta, Glomma River basin, Norway', in Gooch E, Rieu-Clarke A, and Stalnacke P (eds.), *Integrating Water Resources Management: Interdisciplinary Methodologies and Strategies in Practice*, London: IWA Publishing, pp 37–48.

Below R, Grover-Kopec E, and Dilley M, 2007, 'Documenting drought-related disasters: A global reassessment', *Journal of Environment and Development*, 16: 328–344.

Berge D, Barton D, Nhung D, and Nesheim I, 2010, 'The Science-Policy-Stakeholder Interface and Environmental Flow' in Gooche G and Stalnacke P (eds.), *Science, Policy and Stakeholders in Water Management: An Integrated Approach to River Basin Management*, London: Earthscan, pp 105–122.

Bissett N, Comins L, and Seal K, 2010, *Tweed Catchment Management Plan*, Tweed Forum Drygrange, Melrose, Roxburghshire, UK.

Biswas A, 2009, 'Water management: some personal reflections', *Water International*, 34(4): 402–408.

Biswas A and Tortajada C, 2010, 'Future Water Governance: Problems and Perspectives', *International Journal of Water Resources Development*, 26(2): 129–139.

Bjornlund H, 2010, *The Competition for Water: Striking a Balance among Social, Environmental, and Economic Needs*, C.D. Howe Institute Commentary: Governance and Public Institutions No. 302, April 2010.

Boak R and Johnson D, 2007, Hydrogeological impact appraisal for groundwater abstractions, Bristol: Environment Agency Science Report Sc040020/SR2, United Kingdom.

BOND, 2003, *Logical Framework Analysis*, Guidance Notes Series No. 4, viewed 7 March 2013, www.gdrc.org/ngo/logical-fa.pdf.

Boulle L, 2001, *Mediation – Skills and Techniques*, Chatswood: Butterworths.

Bourblanc M, Crabbe A, Liefferink D and Wiering M, 2012, 'The marathon of the hare and the tortoise: implementing the EU Water Framework Directive', *Journal of Environmental Planning and Management*, 56:1449–1467. DOI:10.1080/09640568.2012.726197

Boyle A, 2005, *The Practitioner's Certificate in Mediation Course Handbook*, Melbourne: The Institute of Arbitrators and Mediators Australia.

Bradlow D and Salmon S, 2006, *Regulatory Frameworks for Water Resources Management: A Comparative Study*, Washington DC: World Bank.

Brisbane Declaration, 2007, *The Brisbane Declaration. Environmental Flows are Essential for Freshwater Ecosystem Health and Human Well-Being*, Declaration of the 10th International Riversymposium and International Environmental Flows Conference, Brisbane, Australia, 3–6 September 2007.

Brizga S, 2007, 'The first decade of water resource planning in Queensland: What have we learned about environmental flows?' paper presented to *5th Australian Stream Management Conference*. Australian rivers: Making a difference, Albury, NSW, May 2007.

Brodie RS, Baskaran S, Tottenham R, Hostetler S and Ransley T, 2007, *An adaptive management framework for connected groundwater–surface water resources in Australia*, Canberra: Bureau of Rural Sciences.

Brooks D, 2006, 'An operational definition of water demand management', *Water Resources Development*, 22(4): 521–528.

Brownill S and Parker G, 2010, 'Why bother with good works? The relevance of public participation(s) in planning in a post-collaborative era', *Planning Practice & Research*, 25:3, 275–282.

Bunn S and Arthington A, 2002, 'Basic principles and ecological consequences of altered flow regimes for aquatic biodiversity', *Environmental Management* 30: 492–507.

Burdge R, 2004, 'Social impact assessment: Definition and historical trends', in R Burdge (ed.), *The Concepts, Process and Methods of Social Impact Assessment*, Middleton, Wisconsin: Social Ecology Press, pp. 3–11.

Burrell M, Moss P, Petrovic J and Ali A, 2013, *General Purpose Water Accounting Report 2011–2012: Namoi Catchment*, NSW Department of Primary Industries, Sydney.

Charlton R and Dewdney M, 2004, *The Mediator's Handbook: Skills and Strategies for Practitioners*, Sydney: Lawbook Co.

Chevalier J, 2001, *Stakeholder Analysis and Natural Resource Management*, Ottawa: Carleton University.

Chiew FHS, Teng J, Kirono D, Frost AJ, Bathols JM, Vaze J, Viney NR, Young WJ, Hennessy KJ and Cai WJ, 2008, *Climate data for hydrologic scenario modelling across the Murray-Darling Basin*. A report to the Australian Government from the CSIRO Murray-Darling Basin Sustainable Yields Project, Collingwood, Victoria: CSIRO.

Coleman G, 1987, 'Logical framework approach to the monitoring and evaluation of agricultural and rural development projects', Project Appraisal, 2:4, 251–259.

Commonwealth of Australia 2004, *Intergovernmental Agreement on a National Water Initiative between the Commonwealth of Australia, and the Governments of New South Wales, Victoria, Queensland, South Australia, the Australian Capital Territory and the Northern Territory*, viewed 3 July 2007, http://www.nwc.gov.au/nwi/docs/iga_national_water_initiative.pdf.

Cook N, Schneider G, *River Styles in the Hunter Catchment*, NSW Department of Natural Resources, September 2006.

Coombes P and Barry M, 2008, 'The relative efficiency of water supply catchments and rainwater tanks in cities subject to variable climate and the potential for climate change'. *Australian Journal of Water Resources*, 12 (2), 85–99.

Corvalan C, Hales S and McMichael A, 2005, *Ecosystems and Human Well-being: Health Synthesis: a Report of the Millennium Ecosystem Assessment*, Geneva: WHO (World Health Organisation).

Cox M, James J, Hawke A and Raiber M, 2013, 'Groundwater Visualisation System (GVS): a software framework for integrated display and interrogation of conceptual hydrogeological models, data and time-series animation', *Journal of Hydrology*, 491, 56–72.

Crase L, 2008a, 'Lessons from Australian Water Reform', in Crase L (ed.) *Water Policy in Australia: The Impact of Change and Uncertainty*, Washington, DC, USA: Resources for the Future.

Crase L, 2008b, 'An Introduction to Australian Water Policy', in Crase, L (ed.) *Water Policy in Australia: The Impact of Change and Uncertainty*, Washington, DC, USA: Resources for the Future.

Cullen P, 2004, 'The Knowledge Business of Natural Resource Management', paper presented to Natural Resource Management Ministerial Council, Adelaide, April 2004.

Corkal DR, Diaz H and Sauchyn D, 2011, 'Changing roles in Canadian water management: A case study of agriculture and water in Canada's South Saskatchewan river basin', *International Journal of Water Resources Development*, 27(4): 647–664.

Davidson J, Lockwood M, Curtis A, Stratford E and Griffith R, 2006, *Governance Principles for Regional Natural Resource Management*, Pathways to good practice in regional NRM governance, Canberra: Land and Water Australia.

DEFRA (Department for Environment, Food and Rural Affairs), 2006, *River Basin Planning Guidance*, UK. Viewed 15 August 2011, www.defra.gov.uk.

DEFRA (Department for Environment, Food and Rural Affairs), 2006, *River Basin Planning Guidance*, Defra Publications, available at www.defra.gov.au (accessed 30 March 2013).

DNRE (Victoria Department of Natural Resources and Environment) 2002, *FLOWS – A method for determining environmental water requirements in Victoria*, Department of Natural Resources and Environment.

DNRETAS (NT Department of Natural Resources, Environment, Tourism and Arts and Sports), 2011 *Draft Mataranka (Tindal Aquifer) Water Allocation Plan*, Darwin, NT.

DNRW (Department of Natural Resources and Water), 2007, *Queensland community waterway monitoring manual*, Brisbane: Queensland Government.

DoW (WA Department of Water), 2006, *Ord River water management plan*, Water Resource Allocation and Planning Series Report no. WRAP 15, DoW, Perth, WA.

Doyle S, 2009, 'Aging Urban Infrastructure', *The Water Chronicles*, viewed 24 November 2013, http://www.water.ca/urban.asp.

DPIW (Tasmania Department of Primary Industries and Water), 2009, Generic Principles for Water Management Planning, Water Resources Policy #2005/1. February 2009.

DPW (Department of Public Works), 2010, Official use of social media guideline, ICT Policy and coordination office, DPW, Queensland government, December 2010.

DPIPWE (Tasmania Department of Primary Industries, Parks, Water and Environment), 2012. *Draft Ringarooma River Catchment Water Management Plan.* Water and Marine Resources Division, Department of Primary Industries, Parks, Water and Environment, Hobart.

DPIWE (Tasmania Department of Primary Industries, Water and the Environment), 2005, *River Clyde Water Management Plan*, Department of Primary Industries, Water and the Environment, Hobart.

Duda A and La Roche D, 1997, 'Joint institutional arrangements for addressing transboundary water resources issues – lessons for the GEF', *Natural Resources Forum*, 21(2): 127–137.

DWA (Department of Water Affairs, Republic of South Africa), 2012, *Proposed National Water Resource Strategy 2 {NWRS2}: Summary – Management Water for an Equitable and Sustainable Future*, July 2012, Pretoria.

DWA (Department of Water Affairs, Republic of South Africa), 2013, *National Water Resource Strategy 2 {NWRS2}: Water for an Equitable and Sustainable Future*, June 2013, Pretoria.

DWAF (Department of Water Affairs and Forestry, Republic of South Africa), 2007, *Guidelines for the Development of Catchment Management Strategies: Towards Equity, Efficiency and Sustainability in Water Resources Management*, first edition, by SR Pollard, D du Toit, Y Reddy and T Tlou, Pretoria.

Dyson M, Bergkamp G, Scanlon J. (eds.), *2003, Flow: The Essentials of Environmental Flows*, Gland, Switzerland: IUCN.

EC (European Commission), 2012, *A Commission Report to the European Parliament and the Council on the Implementation of the Water Framework Directive* (2000/60/EC) *River Basin Management Plans*, Brussels: European Commission.

EC staff (European Commission), 2012, Commission Staff Working Document European Overview (1/2) Accompanying the document *Report from the Commission to the Commission to the European Parliament and the Council on the Implementation of the Water Framework Directive (2000/60/EC) River Basin Management Plans*, Brussels: European Commission.

EC (European Communities), 2009, *Common Implementation Strategy for the Water Framework Directive* (2000/60/EC}, Guidance Document on Exemptions to the Environmental Objectives, Guidance Document No. 20, Luxembourg: European Communities.

EC, 2000, EU Water Framework Directive (WFD) Directive 2000/60/EC of the European Parliament and of the Council of 23 October 2000 establishing a framework for the community action in the field of water policy. Official Journal of the European Communities, (L 327), pp. 1–72. viewed 2 March 2013, http://ec.europa.eu/environment/water/water-framework/index_en.html.

Ehrmann J and Stinson B, 2006, *Joint Fact-Finding and the Use of Technical Experts*, Massachusetts Institute of Technology, viewed 30 October 2006, <web.mit.edu/dusp/epp/music/>.

FAO, 2011, *Global Food Losses and Food Waste – Extent, Causes and Prevention*, Rome: Food and Agriculture Organization of the United Nations.

FAO, 2012, *Water News: Climate Change and Water*, viewed 22 November 2012, http://www.fao.org/nr/water/news/clim-change.html.

Faurès J, Bernardi M and Gommes R, 2010, 'There is no such thing as an average: How farmers manage uncertainty related to climate and other factors', *International Journal of Water Resources Development*, 26:4, 523–542.

Ferreyra C, de Loe R and Kreutzwiser R, 2009, 'Imagined communities, contested watersheds: Challenges to integrated water resources management in agricultural areas', *Journal of Rural Studies*, 24: 304–321.

Fiedler K and Doll P, 2007, 'Global modelling of continental water storage changes – sensitivity to different climate data sets', *Advanced Geoscience*, 11: 63–68.

Fisher R, Ury W and Patton B, 1991, *Getting to Yes: Negotiating an Agreement without Giving In*, 2nd edition, Sydney: Random House.

Forrester J and Cinderby S, 2011, *A Guide to using Community Mapping and Participatory-GIS*, prepared for the Managing Borderlands project. Stockholm Environment Institute and Tweed Forum, viewed 20 November 2013, http://www.tweedforum.org/research/Borderlands_Community_Mapping_Guide_.pdf.

García-Vera M, 2013, 'The application of hydrological planning as a climate change adaptation tool in the Ebro basin', *International Journal of Water Resources Development*, 29:2, 219–236.

GHD, Hamstead Consulting and O'Keefe V, 2011, *A framework for managing and developing groundwater trading*, Waterlines report, Canberra: National Water Commission.

Gilman P, Pochat V and Dinar A, 2008, 'Whither la Plata? Assessing the state of transboundary water resource cooperation in the basin', *Natural Resources Forum*, 32: 203–214.

Giordano M and Wolf A, 2003, 'Sharing waters: Post-Rio international water management', *Natural Resources Forum*, 27: 163–171.

Gooch G and Stalnacke P (eds.), 2010, *Science, Policy and Stakeholders in Water Management: An Integrated Approach to River Basin Management*, London: Earthscan.

Gooch G, Allan A, Rieu-Clarke A and Baggett S, 2010, 'The Science-Policy-Stakeholder Interface in Sustainable Water Management: Creating Interactive Participatory Scenarios together with Stakeholders', in Gooch E, Rieu-Clarke A and Stalnacke P (eds.), *Integrating Water Resources Management: Interdisciplinary Methodologies and Strategies in Practice*, London: IWA Publishing, pp 51–65.

Gordon N, McMahon T, Finlayson B, Gippel C and Nathan R, 2004, *Stream Hydrology: An Introduction for Ecologists*, 2nd edition, Chichester, England: John Wiley and Sons.

Gowlland Gualtieri A, 2007, *South Africa's Water Law and Policy Framework: Implications for the Right to Water*, IELRC Working Paper 2007–03, International Environmental Law Research Centre, Geneva, Switzerland, viewed 10 May 2013, http://www.ielrc.org/content/w0703. pdf.

Grafton Q, PIttock J, Davis R, Williams J, Fu G, Warburton M, Udall B, McKenzie R, Yu X, Che N, Connell D, Jian Q, Kompas T, Lynch A, Norris R, Possingham H, and Quiggin J, 2013, 'Global Insights about Water, Climate Change and Governance', *Nature Climate Change*, 3: 315–321.

Graham J, Amos B, Plumtree T, 2003, *Principles for Good Governance in the 21st Century*, Policy Brief No. 15, August 2003, Institute on Governance, viewed 30 July 2013, www.iog.ca.

Green C and Fernandez-Bilbao A, 2006, 'Implementing the Water Framework Directive: How to define a competent authority', *Journal of Contemporary Water Research & Education*, 135: 65–73.

GRI (Global Reporting Initiative), 2005, Sector supplement for public agencies, Pilot version 1, March 2005, Amsterdam. Available from http://www.globalreporting. org/ ReportingFramework/SectorSupplements/PublicAgency/ (accessed 27 October 2009).

Grimble R, 1998, *Stakeholder Methodologies in Natural Resource Management, Socioeconomic Methodologies. Best Practice Guidelines,* Chatham, UK: Natural Resources Institute.

Gurría A, 2009, 'Sustainably managing water: challenges and responses', *Water International*, 34(4): 396- 401.

GWP (Global Water Partnership), 2006, *Technical Brief 3 Monitoring and evaluation indicators for IWRM strategies and Plans*, Stockholm, Sweden: Global Water Partnership.

GWP (Global Water Partnership), 2008, *Toolbox for IWRM* (viewed 2 March 2013) http:// www.gwptoolbox.org/index.php?option=com_content&view=article&id=6&Itemid=4.

GWP (Global Water Partnership), 2012, *Technical Brief 3: Monitoring and evaluation indicators for IWRM strategies and plans*, viewed 1 April 2013, http://www.gwp.org/Global/ GWP-CACENA_Files/en/pdf/tec_brief_3_monitoring.pdf.

GWP TAC (Global Water Partnership Technical Advisory Committee), 2000, *Integrated Water Resources Management,* TAC Background Paper No. 4, Stockholm: GWP.

Hajkowicz S, Young M, Wheeler S, Hatton MacDonald D and Young D, 2000, *Supporting Decisions – Understanding Natural Resource Management Assessment Techniques* – A report to the Land and Water Resources Research and Development Corporation, Collingwood, Vic, June 2000.

Hamstead M, 2009, *Improving environmental sustainability in water planning,* National Water Commission, Waterlines Report Series No20, September 2009, Canberra: National Water Commission.

Hamstead M, 2010, *Alignment of water planning and catchment planning*, Waterlines report, Canberra: National Water Commission.

Hamstead M, 2013, *Proposed Template For Water Sharing Plan Evaluation Reports*, Report to the NSW Office of Water, February 2013, unpublished.

Hamstead M, Baldwin C and O'Keefe V, 2008a, *Water Allocation Planning in Australia – Current Practices and Lessons Learned*, Waterlines Report Series No 6, April 2008, Canberra: National Water Commission.

Hamstead M, Baldwin C and O'Keefe V, 2008b, *An Approach to Water Allocation Planning in Northern Territory*, Northern Territory Department of Natural Resources, Environment, The Arts and Sport, October 2008.

Harrington G and Currie D, 2008, *Conceptual Model Report for Wesley Vale*, Report under the Development of Models for Tasmanian Groundwater Resources Project, DPIW Tasmania, December 2008.

Hassall and Assoc P/L, Ross H and Maher M, 2003, *Profiling – Social and Economic Context: Social Impact Assessment of Possible Increased Environmental Flow Allocations to the River Murray System*, Stage 1, Volume 2, Sydney: Murray-Darling Basin Commission.

Hausler S and Fenton M, 2000, *Burnett Basin WAMP: social assessment report (preliminary draft)*, Queensland Dept. of Natural Resources, Brisbane Australia 2000

Heiland S, 2009, 'Key elements for good governance in water management', *Rural Development News*, 1/2009: 54–57.

Hogan, C 2002, *Understanding Facilitation: Theory and Principles*, London: Kogan Page Ltd.

Horne J, 2013, 'Australian water policy in a climate change context: Some reflections', *International Journal of Water Resources Development*, 29(2): 137–151.

IACSEA (Independent Advisory Committee on Socio-economic Analysis), 1998, *Socio-economic Assessment Guidelines for River, Groundwater and Water Management Committees*, NSW State Government.

IAIA (International Association for Impact Assessment), 2003, *Social Impact Assessment: International Principles*, Fargo, ND: IAIA.

IAP2 (International Association for Public Participation), 2007, *Practitioner tools* (viewed 20 November 2013) http://www.iap2.org/displaycommon.cfm?an=5.

IAP2 (International Association for Public Participation), 2007, *IAP2 Core Values* and *IAP2 Spectrum*, www.iap2.org (accessed November 2013).

ICPR (International Commission for the Protection of the Rhine), 2009, Internationally Coordinated management Plan for the International River Basin District of the Rhine, December 2009, Koblenz, Germany.

ICEWaRM, 2005, *Gaps in skills, training and education in water management: A preliminary report for the National Water Commission*, Adelaide, SA, October 2005.

IPCC (Intergovernmental Panel on Climate Change), 2008, *Technical paper on climate change and water*. June 2008, p.4 (viewed 19 October 2009) http://www.ipcc.ch/pdf/technical-papers/climate-changewater-en.pdf.

IUCN, 2004, *Managing Evaluations: A Guide for IUCN Programme and Project Managers*, Gland, Switzerland and Cambridge, UK: IUCN.

Jackson S, 2008, 'Recognition of Indigenous interests in Australian water resource management, with particular reference to environmental flow assessment', *Geography Compass*, 2/3: 874–898.

Jackson S and Morrison J, 2007, 'Indigenous perspectives in water management, reforms and implementation', in K Hussey and K Dovers (eds.), *Managing Water for Australia*, CSIRO Publishing, Collingwood, Vic, pp. 23–42.

Jackson S and Altman J, 2009, 'Indigenous rights and water policy: Perspectives from tropical Northern Australia', *Australian Indigenous Law Review*, 13 (1): 37–48.

Jackson S and Barber M, 2013, 'Recognition of indigenous water values in Australia's

Northern Territory: current progress and ongoing challenges for social justice in water planning', *Planning Theory and Practice,* DOI: 10.1080/14649357.2013.845684.

Jackson S, Tan P-L, Mooney C, Hoverman S and White I, 2012. 'Principles and guidelines for good practice in Indigenous engagement in water planning', *Journal of Hydrology,* 474: 57–65.

Jønch-Clausen T and Fugl J, 2001, 'Firming up the conceptual basis of integrated water resources management', *International Journal of Water Resources Development,* 17(4): 501–510.

Karlsson-Vinkhuyzen S, 2012, 'From Rio to Rio via Johannesburg: Integrating institutions across governance levels in sustainable development deliberations', *Natural Resources Forum,* 36: 3–15.

Keskinen M and Varis O, 2012, 'Institutional cooperation at a basin level: For what, by whom? Lessons learned from Cambodia's Tonle Sap Lake', *Natural Resources Forum,* 36: 50–60.

King J, Brown C and Sabet H, 2003, 'A scenario-based holistic approach to environmental flow assessments for rivers', *River Research Applications,* 19: 619–639. doi: 10.1002/rra.709.

Klock J and Sjah T, 2011, 'Farmer water management strategies for dry season water shortages in Central Lombok, Indonesia', *Natural Resources,* 2: 114–124.

Landers DH and Nahlik AM, 2013, *Final Ecosystem Goods and Services Classification System* (FEGS-CS). EPA/600/R–13/ORD–004914. Washington, DC: U.S. Environmental Protection Agency, Office of Research and Development.

Laurian L and Shaw M, 2009, 'Evaluation of public participation: The practices of certified planners', *Journal of Planning Education and Research,* 28(3): 293–309.

Le Blanc D, 2012, 'Introduction: Special issue on institutions for sustainable development', *Natural Resources Forum,* 36:1–2.

Lenton R and Muller M (eds.), 2009, *Integrated Water Resources Management in Practice: Better Water Management for Development,* London, UK: Earthscan.

Liguori T, 2009, 'The principle of good faith in the Argentina–Uruguay pulp mills dispute', *The Journal of Water Law.* Special issue: Promoting water for all, 20 (2/3):70–75.

Lockwood M, Davidson J, Curtis A, Stratford E and Griffith R, 2010, 'Governance principles for natural resource management', *Society & Natural Resources: An International Journal,* 23(10): 986–1001.

Loucks D P and van Beek E, 2005, *Water Resources Systems Planning and Management: An Introduction to Methods, Models and Applications,* Paris: UNESCO.

McCaffrey S, 2003, 'The need for flexibility in freshwater treaty regimes', *Natural Resources Forum,* 27: 156–162.

McColl J and Young M, 2005, *Managing change: Australian structural adjustment lessons for water,* Report no. 16/05, Acton ACT: CSIRO Land and Water.

Mackenzie J and Bodsworth P, 2009, *Capacities and needs of water planners in Australia,* Water Planning Tools Milestone Report, Griffith University, CSIRO and National Water Commission, available at http://www.waterplanning.org.au (accessed August 2011).

Mackenzie J, Nolan S and Whelan J, 2009, *Collaborative Water Planning: Guide to Monitoring and Evaluating Public Participation,* Volume 5, TRaCK, Charles Darwin University, Darwin, September 2009.

Marsden Jacobs Associates, 2007a, *The cost-effectiveness of rainwater tanks in Australia.* Prepared for the National Water Commission, Canberra, March 2007.

Marsden Jacobs Associates, 2007b, *The economics of rainwater tanks and alternative water supply options.* Prepared for the Australian Conservation Foundation, Nature Conservation Council (NSW) and Environment Victoria, Melbourne, April 2007.

Martin M and Hill J, 2006, 'Research study into access to recycled water: impediments to recycled water investment', *Australian Journal of Water Resources,* 10 (3): 277–282.

Mati B, 2007, *100 Ways to manage water for smallholder agriculture in Eastern and Southern Africa*: A Compendium of Technologies and Practices, SWMET Working Paper 13, Improved Management in Eastern and Southern Africa (IMAWESA), Nairobi, Kenya, March 2007.

Matthews N, 2012, 'Water grabbing in the Mekong basin: An analysis of the winners and losers of Thailand's hydropower development in Lao PDR', *Water Alternatives*, 5(2): 392–411.

MEA, 2003, *Ecosystems and human well-being: a framework for assessment*, Millennium Ecosystem Assessment, Washington, DC: Island Press.

Mehta L, Veldwisch G and Franco J, 2012, 'Introduction to the Special Issue: Water grabbing? Focus on the (re)appropriation of finite water resources', *Water Alternatives*, 5(2): 193–207.

Menz M, Dixon K and Hobbs R, 2013, 'Hurdles and opportunities for landscape-scale restoration', *Science*, 336: 526–527.

Michels A and De Graaf L, 2010, 'Examining citizen participation: Local participatory policy making and democracy', *Local Government Studies*, 36 (4): 477–491.

Miner M, Patankar G, Gamkhar S and Eaton D, 2009, 'Water sharing between India and Pakistan: a critical evaluation of the Indus Water Treaty', *Water International*, 34(2): 204–216.

Mirza S, 2007, *Danger Ahead: The Coming Collapse of Canada's Municipal Infrastructure: A Report for the Federation of Canadian Municipalities*, Federation of Canadian Municipalities, Ottawa, Canada, viewed 24 November 2013, www.fcm.ca.

Molden D and Sakthivadivel R, 1999, 'Water accounting to assess use and productivity of water', *International Journal of Water Resources Development*, 15(1–2), 55–71.

Mooney C, Baldwin C, Tan P-L and Mackenzie J, 2012, 'Transparency and trade-offs in water planning', *Journal of Hydrology*, 474: 66–73.

Mooney C and Tan P-L, 2012, 'South Australia's River Murray: Social and cultural values in water planning', *Journal of Hydrology*, 474, 29–37

Moss T, 2004, 'The governance of land use in river basins: prospects for overcoming problems of institutional interplay with the EU Water Framework Direction', *Land Use Policy*, 21: 85–94.

Mostert E, 2003, 'The challenge of public participation', *Water Policy*, 5(2):178–197.

Murray A and Seddon S, 2008, *Preliminary Water Industry Interaction with Universities Survey*, Artarmon, NSW: Water Industry Capacity Development.

Nancarrow B, Leviston Z, Tucker D, Greenhill M, Price J and Dzidic P, 2007, *Community acceptability of the indirect potable use of purified recycled water in South East Queensland and preferences for alternative water sources: a baseline measure*, for Urban Water Security Research Alliance, Technical Report No 1, CSIRO, November 2007.

NLWRA (National Land and Water Resource Audit), 2000, *Australian Dryland Salinity Assessment 2000. Extent, impacts, processes, monitoring and management options*, Canberra: National Land and Water Resource Audit.

NSW Government, 2002, *Water Policy Advisory Notes for Water Management Committees*, Sydney, NSW, viewed 8 November 2013, http://www.water.nsw.gov.au/Water-management/Water-sharing-plans/planning-process/default.aspx.

NSW Government, 2003, *Water Sharing Plan for the Macquarie and Cudgegong Regulated Rivers 2003*, Sydney, NSW.

NSW Government, 2010, *Water Sharing Plan for the Peel Valley Regulated, Unregulated, Alluvium and Fractured Rock Water Sources*, viewed 30 July 2013, http://www.legislation.nsw.gov.au/viewtop/inforce/subordleg+134+2010+cd+0+N/

NSW Government, 2014, *Water Management Act 2000*, as at 1 January 2014, http://www.legislation.nsw.gov.au/viewtop/inforce/act+92+2000+FIRST+0+N/ (accessed 7 April 2014)

NSW OofW (NSW Office of Water), 2011, *Macro water sharing plans – the approach for unregulated rivers – A report to assist community consultation*, Second edition, NSW Government Office Of Water, Sydney, NSW. August 2011.

NT (Northern Territory) Government, 2009, *Water Allocation Plan for the Tindall Limestone Aquifer, Katherine 2009–2019*, Darwin, NT.

NWC (National Water Commission), 2009, *Australian Water Reform 2009: Second biennial assessment of progress in implementation of the National Water Initiative*, Canberra: National Water Commission.

NWC (National Water Commission), 2010, *The impacts of water trading in the southern Murray-Darling Basin: an economic, social and environmental assessment*, Canberra: National Water Commission.

NWC (National Water Commission), 2011a, *Environmental Water Management Assessment and Reporting Framework*, Canberra: National Water Commission.

NWC (National Water Commission), 2011b, *Water markets in Australia: a short history*, Canberra: National Water Commission.

NWC (National Water Commission), 2011c, *The National Water Initiative – securing Australia's water future: 2011 assessment*, Canberra: National Water Commission.

NWC (National Water Commission), 2013, *Water management and pathways to sustainable levels of extraction: issues paper*, Canberra: National Water Commission.

Ostrom E, 1992, *Crafting Institutions for Self-Governing Irrigation Systems*, California: ICS Press.

Ostrom E, 2005, *Understanding Institutional Diversity*, Princeton, NJ: Princeton University Press.

Ostrom E and Nagendra H, 2007, 'Tenure alone is not sufficient: monitoring is essential', *Environmental Economics and Policy Studies*, 8 (3):178–199.

Ostrom E, Berger J, Field C, Norgaard R and Policansky D, 1999, 'Revisiting the Commons: Local Lessons, Global Challenges', *Science*, 284: 278–282.

Pahl-Wostl C, Kabat P, Moltgen J (eds.), 2008, *Adaptive and Integrated Water Management: Coping with Complexity and Uncertainty*, Berlin: Springer.

Paranjape S, Joy K, Mansi S, Latha N and Mollinga P, 2010, 'Integrating tanks into the larger waterscape in the Tungabhadra', in Gooch E, Rieu-Clarke A and Stalnacke P (eds.), *Integrating Water Resources Management: Interdisciplinary Methodologies and Strategies in Practice*, London: IWA Publishing, pp. 93–104.

Parris K, 2011, 'Improving the information base to better guide water resource management decision making', *International Journal of Water Resources Development*, 27(4): 625–632.

Petit O and Baron C, 2009, 'Integrated water resources management: From general principles to its implementation by the state. The case of Burkina Faso', *National Resources Forum,* 33: 49–59.

Plant R, Taylor C, Hamstead M and Prior T, 2012, *Recognising the broader benefits of aquatic systems in water planning: an ecosystem services approach*, Waterlines report, National Water Commission, August 2012 Canberra, Australia.

Po M, Kaercher J and Nancarrow B, 2003, *Literature review of factors influencing public perceptions of water reuse*, Technical Report 54/03 CSIRO Land and Water.

Poff NL; Richter, BD; Arthington, A; Bunn, S; Naiman, RJ; Kendy, E; Acreman, M; Apse, C; Bledsoe, BP; Freeman, MC; Henriksen, J; Jacobson, RB; Kennen, JG.; Merritt, DM; O'Keeffe, JH; Olden, JD; Rogers K; Tharme RE and Warner A, 2010, 'The ecological limits of hydrologic alteration (ELOHA): a new framework for developing regional environmental flow standards', *Freshwater Biology*, 55 (1), pp. 147–170.

Prell C, Hubacek K and Reed M, 2009, 'Stakeholder Analysis and Social Network Analysis in Natural Resource Management', *Society & Natural Resources*, 22(6): 501–518.

Proctor W, 2009, 'Multi-Criteria Analysis', in Argyrous G (ed.), *A Practical Guide: Evidence for Policy and Decision Making*, Sydney: University of New South Wales Press.

Proctor W, 2010, *DMCE Case Studies Overview*. Unpublished paper.

QG (Queensland Government), 2000, *Water allocation and management plan Burnett Basin: draft condition and trend report*, Brisbane: Queensland Department of Natural Resources.

QG (Queensland Government), 2006, *Burnett Basin ROP*, Brisbane: Queensland Department of Natural Resources.

QG (Queensland Government), 2009, *South East Queensland Infrastructure Plan and Program: 2009–2026*, Department of Infrastructure and Planning, Brisbane, Australia.

Quinlan R, Arthington A and Brizga S, 2004, *Benchmarking, a "Top-Down" Methodology for Assessing the Environmental Flow Requirements of Queensland Rivers*, Queensland Conservation Council, Brisbane, April 2004.

QWC (Queensland Water Commission), 2009a, *Southeast Queensland Water strategy – draft*. QWC, December 2009, Brisbane.

QWC (Queensland Water Commission), 2009b, *Permanent water conservation measures*, viewed 14 October 2009, http://www.qwc.qld.gov.au/Permanent _Water_Conservation_Measures.

QWI (Queensland Water Infrastructure Pty Ltd), 2007, Traveston Crossing Dam project, project update number four, October 2007, viewed 12 October 2009, http://qldwi.com.au/LinkClick.aspx?fileticket_Q7Om%2BCQAe3o%3D&tabid_59&mid_477.

Rahaman M, 2009, 'Principles of transboundary water resources management and Ganges treaties: An analysis', *International Journal of Water Resources Development*, 25(1): 159–173.

Rahaman M and Varis O, 2005, 'Integrated water resources management: evolution, prospects and future challenges', *Sustainability: Science, Practice, & Policy*, 1(1):15–21.

Ramsar, 2013, *The Ramsar Convention on Wetlands*, viewed 2 March 2013, http://www.ramsar.org/cda/en/ramsar-european-rs-omeindex/main/ramsar/1%5E26097_4000_0__.

Reid M, Cheng X, Banks E, Jankowski J, Jolly ID, Kumar P, Lovell D, Mitchell M, Mudd G, Richardson S, Silburn M and Werner A, 2009, *Catalogue of conceptual models for ground-water–stream interaction*, eWater Technical Report. eWater Cooperative Research Centre, Canberra. viewed 20 November 2013, http://ewatercrc.com.au/reports/Reid_et_al–2009-Model_Catalogue.pdf.

Republic of Kenya, 2009, *Tana Water Catchment Area Management Strategy*, Nairobi, Kenya: Water Resource Management Authority.

Rieu-Clarke A and Loures FR, 2009, 'Still not in Force: Should States Support the 1997 UN Watercourses Convention?', *Review of European Community and International Environmental Law* 185, 18(2): 185–197.

Ross H, Buchy M and Proctor W 2002, 'Laying down the ladder: A typology of public participation in Australian natural resource management', *Australian Journal of Environmental Management*, 9(4): 205–17.

Roughley A and Williams S, 2007, *The Engagement of Indigenous Australians in Natural Resource Management: Key Findings and Outcomes from Land and Water Australia funded research and broader literature*, Land and Water Australia, November 2007.

Rowe G and Frewer L, 2000, 'Public participation methods: A framework for evaluation', *Science, Technology & Human Values*, 25(1): 3–29

Rowe G and Frewer L, 2005, 'A Typology of Public Engagement Mechanisms', *Science Technology Human Values*, 30: 251–290.

Salgado P, 2009, 'Participative multi criteria analysis for the evaluation of water governance alternatives: A case in the Costa del Sol (Malaga)', *Ecological Economics*, 68: 990–1005.

Salman S, 2007, 'The Helsinki Rules, the UN Watercourses Convention and the Berlin Rules:

Perspectives on International Water Law', *International Journal of Water Resources Development*, 23(4): 625–640.

Sarker A, Baldwin C and Ross H, 2009, 'Managing groundwater as a common-pool resource: an Australian case study', *Water Policy*, 11: 598–614.

Sarkissian W, Cook A and Walsh K, 1999, *Community Participation in Practice: A Practical Guide*, Murdoch, WA: Murdoch University.

Sarkissian W, Hirst A, Stenberg B and Walton S, 2003, *New Directions: Community Participation in Practice*, Murdoch, WA: The Institute for Sustainability and Technology Policy, Murdoch University.

Seckler D, Barker R and Amarasinghe U, 1999, 'Water scarcity in the twenty-first century', *International Journal of Water Resources Development*, 15(1–2): 29–42.

SENRMB, 2011, *Guide to the development and contents of the 2009 Padthaway water allocation plan*, South East Natural Resources Management Board, Government of South Australia. April 2011.

SEQHWP (South East Queensland Healthy Waterways Partnership), 2010, *Environmental Health Monitoring Program 2008–09 Annual Technical Report Executive Summary*, Brisbane Qld, available at www.healthywaterways.org.

Sidaway R, 2005, *Resolving Environmental Disputes: From Conflict to Consensus*, London: Earthscan.

Singh I, Flavel N and Bari M, 2009, *Coopers Creek Water Sharing Plan – Socioeconomic assessment of changes to the flow rules*, NSW Government, Department of Water and Energy, Sydney, July 2009.

SjahT, 2007, 'Managing production risk of water on new cropping lands in Lombok, Indonesia', in Klock J, and Sjah T (eds.), *Water Management in Lombok, Indonesia: Challenges and Solutions*, Mataram, Indonesia: Mataram University Press.

SKM (Sinclair Knight Merz), 2001, *Environmental Water Requirements of Groundwater Dependent Ecosystems*, Environmental Flows Initiative Technical Report Number 2, Commonwealth of Australia, Canberra.

SKM (Sinclair Knight Merz), 2011, *National framework for integrated management of connected groundwater and surface water systems*, *Waterlines report,* Canberra: National Water Commission,.

Sosa M and Zwarteveen M, 2012, 'Exploring the politics of water grabbing: The case of large mining operations in the Peruvian Andes', *Water Alternatives*, 5(2): 360–375.

South Australian Government, 2001, *Water Allocation Plan for the Padthaway Prescribed Wells Area 2001*, South East Catchment Management Board, October 2001.

South Australian Government, 2009, *Water Allocation Plan for the Padthaway Prescribed Wells Area 2009*, South East Catchment Management Board, April 2009.

South Australian Government, 2011, *Guide to the development and contents of the 2009 Water Allocation Plan for the Padthaway Prescribed Wells Area*, April 2011.

Southeast Asia Rivers Network, AotP, Pak Mun Dam Affected People, Thailand, 2004, *The Return of Fish, River Ecology and Local Livelihoods of the Mun River: A Thai Baan (Villagers) Research*, Chiang Mai, Thailand: Southeast Asia Rivers Network.

Stalnacke P, Gooch G and Rieu-Clarke A, 2010, 'STRIVER – Overall findings' in Gooch E, Rieu-Clarke A, and Stalnacke P (eds.), *Integrating Water Resources Management: Interdisciplinary Methodologies and Strategies in Practice*, London: IWA Publishing, pp. 151–160.

Susskind L, McKearnan S and Thomas-Larmer J (eds.), 1999, *Consensus Building Handbook: A Comprehensive Guide to Reaching Agreement*, Thousand Oaks, CA: Sage Publications.

Swedish Environmental Advisory Council, 2009, *Resilience and Sustainable Development*, Stockholm: Ministry of the Environment,

Syme G, Nancarrow B, McCreddin J, 1999, 'Defining the components of fairness in the allocation of water to environmental and human uses', *Journal of Environmental Management*, 57: 51–70.

Tan P-L, Baldwin C, White I, Burry K, 2012, 'Water planning in the Condamine Alluvium, Queensland: Sharing information and eliciting views in a context of overallocation', *Journal of Hydrology*, 474: 38- 46.

Tan P-L., Mooney C, White I, Hoverman S, Mackenzie J, Burry K, Baldwin C, Bowmer K, Jackson S, Ayre M, and George D, 2010, *Tools for water planning: lessons, gaps and adoption, Waterlines Series Report No 37*, Canberra: National Water Commission, Available at http://www.nwc.gov.au/www/html/2978-waterlines-37.asp?intSiteID=1

Team Technologies (Middleburg, VA), 2005, *The Logframe Handbook: A Logical Framework Approach to Project Cycle Management*. Washington, DC: World Bank. http://documents.worldbank.org/curated/en/2005/01/5846691/logframe-handbook-logical-framework-approach-project-cycle-management

Tharme R, 2003, 'A global perspective on environmental flow assessment: emerging trends in the development and application of environmental flow methodologies for rivers', *River Research and Applications*, 19: 397–441.

Transform, Department of Forestry, WNT Government, Lombok Timur Government and WWF Indonesia Nusa Tenggara, 2005, *Karakteristik sub daerah aliran sungai (DAS) Pohgading Sunggen dan Labuhan Lombok di Kabupaten Lombok Timur* (Characteristics of the sub water catchments of Pohgading Sunggen and Labuhan Lombok in East Lombok), Mataram: Lombok Timur Government.

Turner A, Hausler G, Carrard N, Kazaglis A, White S, Hughes A and Johnson T, 2007, *Review of water supply-demand options for South East Queensland* for Mary River Council of Mayors, Institute for Sustainable Futures, Sydney February 2007.

TVA (Tennessee Valley Authority), 2013, *Tennessee Valley Authority, Managing River System Flows*, accessible at http://www.tva.com/river/lakeinfo/systemwide.htm, (accessed 25 November 2013).

Twyford V, Waters S, Hardy M, Dengate J, 2006, *Beyond Public Meetings: Connecting Community Engagement with Decision-making*, Wollongong, Australia: Vivien Twyford Communication Pty Ltd,.

Uhlmann V, 2004, *An approach to sustainability management for water utilities*, PhD thesis, University of Queensland, Brisbane.

University of Wisconsin-Extension, 2003, *Enhancing Program Performance with Logic Models*, viewed 6 February 2013, http://www.uwex.edu/ces/lmcourse/.

UN General Assembly, 1997, *Convention on the Law of the Non-navigational Uses of International Watercourses*, General Assembly resolution 51/229, annex', *Official Records of the General Assembly, Fifty-first Session, Supplement No. 49* (A/51/49). Not yet in force.

UN (United Nations), 1992, *Earth Summit: Agenda 21: The United Nations Programme of Action from Rio*, United Nations (accessed on 7 April 2014 from http://www.unep.org/Documents. Multilingual/Default.asp?documentid=52).

UNDP, 2000, *Millennium Development Goals*, New York: United Nations Development Programme.

UNEP, 2006, *Global International Waters Assessment, Challenges to International Waters – Regional Assessments in a Global Perspective*, Nairobi: UNEP.

UNEP, 2007, *Dams and Development: Relevant Practices for Improved Decision-making (Compendium)*, Nairobi, Kenya: United Nations Environment Programme.

UNEP, 2009, *Water Security and Ecosystem Services: The Critical Connection*. A contribution to the UN World Water Assessment Programme, Nairobi, March 2009.

UNEP, 2012, *The UN-Water Status Report on the Application of Integrated Approaches to Water Resources Management*, Nairobi, Kenya.

UNEP DEWA/GRID-Europe, 2005, Freshwater consumption in Europe. Available from http://www.grid.unep.ch/product/publication/freshwater_europe/consumption.php (accessed January 2010), p. 14.

United Nations Statistics Division, 2012, *System of Integrated Environmental-Economic Accounts for Water (SEEA-Water)*, New York: United Nations.

van Koppen B, 2008, 'Redressing inequities of the past from a historical perspective: The case of the Olifants basin, South Africa', *Water SA*, 34 (4) (Special HELP edition) http://www.wrc.org.za

Vanclay F and Social Section of IAIA, 2004, 'International principles for social impact assessment', in R Burdge (ed.) *The Concepts, Process and Methods of Social Impact Assessment*, Middleton, WI, USA: Social Ecology Press, pp. 273–281.

Victorian Government, 2002, *Victorian River Health Strategy*. Department of Natural Resources and Environment, August 2002.

Victorian Government, 2004, *Securing our water future together,* Department of Sustainability and the Environment, Melbourne, June 2004.

Victorian Government, 2006, *Sustainable Water Strategy, Central Region – Draft for community comment,* Department of Sustainability and Environment, Melbourne, April 2006.

Victorian Government, 2008a, *Sustainable Water Strategy, Northern Region – Discussion paper,* Department of Sustainability and Environment, Melbourne, January 2008.

Victorian Government, 2008b, *Sustainable Water Strategy, Northern Region – Draft for community comment,* Department of Sustainability and Environment, Melbourne, October 2008.

Victorian Government, 2009, *Sustainable Water Strategy, Northern Region,* Department of Sustainability and Environment, Melbourne, November 2009.

Walker B, Abel N, Anderies J, and Ryan P, 2009, 'Resilience, adaptability and transformability in the Goulburn-Broken Catchment, Australia', *Ecology and Society*, 14 (1):12 (online).

Warrick R, 2009, 'Using SimCLIM for modelling the impacts of climate extremes in a changing climate: a preliminary case study of household water harvesting in Southeast Queensland'. Paper presented at *18th World IMACS/MODSIM Congress*, Cairns, Australia, 13–17 July 2009. Available from http://mssanz.org.au/modsim09

WBM Oceanics Australia, 2006, *Rainwater tank modelling investigation: South East Queensland regional water supply strategy.* Prepared for MWH Australia PL, Spring Hill.

Werner M and Hauber-Davidson G, 2008, 'An integrated approach to water conservation for large users', *Australian Journal for Water Resources*, 12 (2): 179–186.

West Gippsland Catchment Management Authority, 2005, *West Gippsland River Health Strategy 2005*, Traralgon, Victoria.

White I, 2009, *Decentralised environmental technology adoption: The household experience with rainwater harvesting*, PhD thesis, Griffith University, Brisbane, Queensland.

White R, 2010, *A Review of Current Knowledge: World Water: Resources, Usage, and the Role of Man-made Reservoirs*, revised March 2010, Marlow, UK: Foundation for Water Research, available at http://www.fwr.org/wwtrstrg.pdf.

Winter TC, Judson WH, Franke OL and Alley WM, 1998, *Groundwater and surface water a single resource*, Circular 1139, Denver, CO: U.S. Geological Survey.

Wolf A, Natharius J, Danielson J, Ward B, and Pender J, 1999, 'International river basins of the world', *International Journal of Water Resources Development*, 15(4): 387–427.

Wolf A, Yoffe S, Giordano M, 2003, 'International waters; identifying basins at risk', *Water Policy*, 5: 31–62.

World Commission on Environment and Development, 1987, *Our Common Future-the Brundtland Report,* World Water Council 2012, 'Vision, Mission and Strategy' and 'Background', Available at http://www.worldwatercouncil.org/about-us/vision-mission-strategy/ (accessed 2 March 2013).

Appendix

Typical benefits and services derived from aquatic ecosystems

Presented here are tables of typical ecosystem services provided by water resources, along with a typical associated benefit (note services can contribute to multiple benefits – this is not depicted in the table). Services are divided into three types, 'Provisioning', 'Social and Cultural', and 'Regulating'. The tables are taken from Plant *et al.* (2012), pp. 68–71. They are used with permission.

Table A.1 Provisioning services

Key benefit associated with service	*Service provided by aquatic system*
Water at low or no treatment costs for: • drinking • gardens, parks, sporting facilities, recreational areas, etc. • livestock • irrigation of crops, pasture, etc. • industry use, e.g. processing, cooling, wash-down, dust suppression, power, food, clothing, paper, etc.	Provision of water
Water delivered, available and drained for human use at low or minimal cost (natural water 'infrastructure')	Provision of storage, release, drainage and delivery of water, e.g. via network and arrangement of channels, pools, runs, etc., and associated floodplain areas, maintaining floodplain connectivity and maintaining groundwater–surface water balances and storages through recharge and discharge processes
Foods for human consumption, important source of nutrients, etc.	Provision of animal and plant species that are edible, e.g. fish, crustaceans, shellfish. Includes estuarine and near-shore marine species that are dependent on rivers, etc., for food or lifecycle stages. Includes riparian and floodplain species and provision of bush tucker

Key benefit associated with service	*Service provided by aquatic system*
Non-water/-food products for consumptive industries and businesses	Provision of areas, features and/or conditions (excluding water and food) that may be consumed/harvested, e.g. sand and gravel, aquaculture brood stock, biochemical products, specialist wood products, fuel wood, salt, aquarium fish and plants, species for sports hunting, etc.
Non-water/-food products for non-consumptive industries and businesses	Provision of areas, features and/or conditions (excluding water and food) that are not directly consumed/harvested, e.g. provision of nutrients via sediment deposition for floodplain grazing, stock watering points, provision of attractive areas, flora, fauna, etc., for recreation, tourism, hospitality, health and fitness, four wheel driving, etc.
Non-water/-food products that are non-commercial	Provision of non-commercial materials (excluding water and food) used by people either directly or indirectly, e.g. fuel wood, woven products, ornaments, etc. See also social and cultural services
Genetic resources	Provision of a natural reservoir for biological diversity providing genetic resources that can support colonisation, contribute to maintaining intra-species diversity, and selected breeding. Includes the provision of genetic resources for resistance to pathogens, or tolerance to environmental conditions, and the provision of new medicines
Biodiversity benefits ('existence' benefits)	Provision of flora and fauna. Provision of water, food, habitat, nursery sites, etc., for flora and fauna
Energy	Provision of hydroelectricity, geothermal water, etc., opportunities
Transport	Provision of water-based transport opportunities
Income, employment	Provision of materials and opportunities like those above that facilitate economic activities

Table A.2 Social and cultural services

Key benefit associated with service	Service provided by aquatic system
Amenity and aesthetics – attractive landscapes, scenes, surroundings or atmospheres that people can enjoy looking at, smelling, hearing, sensing or otherwise appreciate or derive pleasure from (e.g. the sun sparkling on the water may generate a feeling of joy, mist over a floodplain may generate a sense of peace, sunset over an estuary may generate a sense of inspiration, and the scenes, sounds and smells along a familiar riverside path may generate a sense of home, etc.)	Provision of areas, features or conditions that facilitate sensory enjoyment or pleasure such as attractive landscapes, scenes and environments that offer different views, smells, sounds, senses, atmospheres, etc. These may display 'typical' features, or may be especially beautiful, unusual or 'iconic' (similar benefits can occur at the domestic level through water for gardens)
Social relations and social cohesion – attractive areas to socialise, e.g. natural areas attractive for family gatherings such as picnics and barbeques, for social activities such as fishing, boating, and walking, and for community events (festivals, concerts, etc.). In regional towns presence of/ views of healthy river pool, water birds, etc. can promote a positive community atmosphere (anecdotal evidence may include less juvenile crime, lower incidence of depression, etc.)	Provision of areas, features or conditions that facilitate relationships among people and/or communities, e.g. provision of attractive natural areas, presence of/views of healthy river pool, water birds, etc.
Mental health – areas, features or conditions that are beneficial to mental health, e.g. views, sounds, smells and other sensory experiences that help people relax, unwind, regain perspective, etc. Related to social relations, amenity, identity and physical health	Provision of areas, features or conditions that facilitate mental health, e.g. attractive riparian area or waterbody, trees mirrored in a still pool, flickers of dappled light reflecting off a waterbody, sounds of flowing water or frogs, smell of an estuary
Physical health – attractive areas to exercise, e.g. attractive and accessible riparian areas may encourage people to walk, run, cycle, etc. Appropriate water quality and amenity may encourage rowing, sailing, windsurfing, swimming, canoeing, etc. Other physical health benefits related to aquatic systems may come from activities including water skiing, fishing, hunting, hiking, crabbing, camping, etc. (see also drinking water)	Provision of areas, features or conditions that facilitate good physical health, e.g. attractive and accessible riparian areas, good water quality, various water flow/ level conditions related to recreational opportunities
Recreation – attractive areas for recreational activities such as those listed above in physical health but also more physically passive activities such as sightseeing, four wheel driving, motor boating, jet skiing, and nature observation such as bird-watching and photography. There are important mental health benefits from recreation	Provision of areas, features, or conditions that provide recreation opportunities

Key benefit associated with service	*Service provided by aquatic system*
Tourism – (in addition to benefits described under physical health and recreation) areas, features, or conditions that are especially unique or scenic and benefit people to undertake activities associated with tourism, e.g. swimming in waterholes, admiring waterfalls, rafting on white water, house-boating on flat scenic water, enjoying rare, diverse or abundant flora and fauna. Tourists particularly benefit from water features near roads in rural areas (e.g. popular for overnight camping)	Provision of areas, features, or conditions for people to undertake activities associated with tourism. Many of these services are covered above in recreation, but to attract tourists the aquatic system might provide areas, features or conditions that are especially unique or scenic, e.g. waterfalls, water holes, gorges, cave stalagmites, white water, or rare, diverse or abundant birds, animals or vegetation
Education and research – areas, features or conditions that benefit education and research ranging from school excursions to investigating genetics for medically beneficial materials and future products	Provision of areas, features or conditions that facilitate education and research, e.g. rare species, typical river geomorphology, opportunities for school projects
Cultural heritage & identity — areas, features or conditions that are important for people's identity and culture (both Indigenous and non-Indigenous). For example, the aquatic system may: • contain historical structures or items that connect people with their past or the region's past, e.g. ruined dwellings, fish traps, bridge timbers. These may also include products derived from the aquatic system or tools for hunting aquatic biota • host species of cultural or religious significance, including family and totem species • provide significant nonmaterial culture such as though folklore, music, customs and traditional knowledge • provide sites of spiritual significance, e.g. for ceremony or burial	Provision of areas, features or conditions that are important for people's identity and culture, e.g. natural sites of spiritual significance, species of cultural or religious significance
Social benefits derived from economic benefits created by aquatic systems, e.g. from income and employment in water related industries, etc.	Provision of water and areas, features and conditions facilitating and contributing to socio-economic opportunities

Table A.3   Regulating services

Key benefit associated with service	Service provided by aquatic system
Good water quality	Role in purifying water via slowing flow, filtering, trapping, buffering and assimilating sediments, nutrients and other contaminants
Reduced risk from pollution, algal blooms and other contaminants	Role (beyond performing water purification processes) in mitigating diffuse and point sources of pollution (e.g. runoff from urban or degraded agricultural landscapes, discharges from sewage treatment plants or industry, intensive agriculture, stormwater drains, etc.), e.g. by: • sediments assimilating and storing contaminants such as heavy metals • diluting pollutants (e.g. high flows diluting and exporting saline water from a river system) • maintaining channel capacity and preventing accumulation of toxic sludges, etc. (e.g. high flows scouring sediment slugs) • shade from riparian vegetation preventing high water temperatures conducive to algal blooms
Bank and shoreline stabilisation and erosion protection	Role in stabilising banks and substrate, retaining soils, and in preventing erosion and slumping, e.g. by waning flows depositing sediment, by roots of riparian and aquatic vegetation, by wetlands and floodplains slowing down flow to mitigate channel erosion, etc.
Fewer pest species and diseases	Role in providing habitat, food, etc. for animals and insects that can control pests and diseases. For example, some frogs and fish that live in aquatic systems reduce the abundance of disease vectors by eating mosquitoes or their larvae. Some aquatic systems provide habitat for predators that control agricultural pests, for example, ibis feeding on grasshoppers.
Fewer weed species	Role in preventing or minimising colonisation and growth of weed species (or species not adapted to water regime at a particular site), e.g. via high-flow levels preventing germination, by flood flows scouring sediment containing seeds or juvenile plants, or native riparian vegetation shading out weed species
Micro, local, regional, etc. atmospheric and climate conditions favourable to people	Role of system in influencing atmospheric and climatic factors at range of scales, e.g. regulating ecological and hydrological processes and cycles and therefore temperatures, evaporation, precipitation, greenhouse gases and other climatic processes via, e.g. water temperatures, shade from aquatic, riparian or floodplain vegetation
Reduced risk from floods and other natural physical hazards	Role in reducing floodwater impacts, for example, reducing peak levels and velocities by wetland and floodplain storage, and vegetation roughness. Role in regulating storm effects, shoreline and riverbank stabilisation, and mitigation of fire intensity and frequency in wet and damp lands.

Key benefit associated with service	*Service provided by aquatic system*
Healthy and productive coastal environments	Role in maintaining coastal processes, e.g. sediment supplied to coastal zone from rivers stabilises shorelines and substrate from the impacts of wave action and currents, and provides nutrients. Freshwater events entering estuarine and coastal areas provide important cues to fish and other species.

Index

For Product Safety Concerns and Information please contact our EU
representative GPSR@taylorandfrancis.com
Taylor & Francis Verlag GmbH, Kaufingerstraße 24, 80331 München, Germany

www.ingramcontent.com/pod-product-compliance
Lightning Source LLC
Chambersburg PA
CBHW070352270326
41926CB00014B/2509